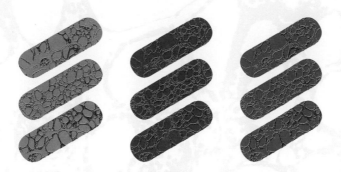

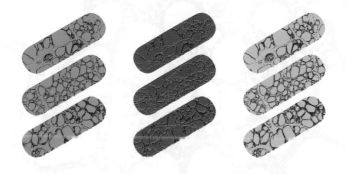 **OXFORD BIOLOGY PRIMERS**

Discover more in the series at
www.oxfordtextbooks.co.uk/obp

Published in partnership with the Royal Society of Biology

BIOCHEMISTRY

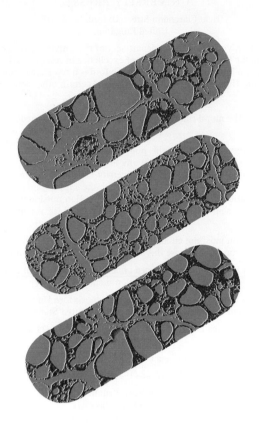

BIOCHEMISTRY

The molecules of life

Richard Bowater, Laura Bowater, and Tom Husband

Edited by Ann Fullick

Editorial board: Ian Harvey, Gill Hickman, and Sue Howarth

OXFORD
UNIVERSITY PRESS

Royal Society of
Biology

OXFORD
UNIVERSITY PRESS

Great Clarendon Street, Oxford, OX2 6DP,
United Kingdom

Oxford University Press is a department of the University of Oxford.
It furthers the University's objective of excellence in research, scholarship,
and education by publishing worldwide. Oxford is a registered trade mark of
Oxford University Press in the UK and in certain other countries

© Oxford University Press 2020

The moral rights of the authors have been asserted

Impression: 1

Public sector information reproduced under Open Government Licence v3.0
(http://www.nationalarchives.gov.uk/doc/open-government-licence/open-government-licence.htm)

Published in the United States of America by Oxford University Press
198 Madison Avenue, New York, NY 10016, United States of America

British Library Cataloguing in Publication Data
Data available

Library of Congress Control Number: 2019952903

ISBN 978-0-19-884839-4

Printed in Great Britain by
Bell & Bain Ltd., Glasgow

PREFACE

Welcome to the Oxford Biology Primers

There has never been a more exciting time to be a biologist. Not only do we understand more about the biological world than ever before, but we're using that understanding in ever more creative and valuable ways.

Our understanding of the way our genes work is being used to explore new ways to treat disease; our understanding of ecosystems is being used to explore more effective ways to protect the diversity of life on Earth; our understanding of plant science is being used to explore more sustainable ways to feed a growing human population.

The repeated use of the word 'explore' here is no accident. The study of biology is, at heart, an exploration. We have written the Oxford Biology Primers to encourage you to explore biology for yourself—to find out more about what scientists at the cutting edge of the subject are researching, and the biological problems they're trying to solve.

Throughout the series, we use a range of features to help you see topics from different perspectives.

Scientific approach panels help you understand a little more about 'how we know what we know'—that is, the research that has been carried out to reveal our current understanding of the science described in the text, and the methods and approaches scientists have used when carrying out that research.

Case studies explore how a particular concept is relevant to our everyday life, or provide an intimate picture of one aspect of the science described.

The bigger picture panels help you think about some of the issues and challenges associated with the topic under discussion—for example, ethical considerations, or wider impacts on society.

More than anything, however, we hope this series will reveal to you, its readers, that biology is awe-inspiring, both in its variety and its intricacy, and that it will drive you forward to explore the subject further for yourself.

ABOUT THE AUTHORS

Richard Bowater

Richard Bowater gained his PhD from the University of London and currently works in the School of Biological Sciences at the University of East Anglia (UEA), where he teaches biochemistry and molecular biology to all levels of university students. Richard has authored many primary publications and reviews that focus on his research interests, and also delivered pedagogical commentaries about teaching biochemical concepts to diverse audiences. He became a Senior Fellow of the Higher Education Academy (SFHEA) in the UK in 2015, is a Fellow of the Royal Society of Biology, and is a member of the Biochemical Society, the Microbiology Society, and the Royal Society of Chemistry. In collaboration with the Biochemical Society, Richard and other colleagues from UEA developed a massive open online course (MOOC) entitled *Biochemistry: Molecules of Life*, which is targeted at fifteen–nineteen-year-old students and is delivered on the FutureLearn platform.

Laura Bowater

Laura Bowater is Professor of Microbiology Education and Engagement at the University of East Anglia. She began her academic career as a microbiologist at the University of Dundee, graduating with an MSc and a PhD in microbial biochemistry. She continued her scientific research at the John Innes Centre in Norwich, UK. Laura has combined her interest in microbiology and the growing problem of antimicrobial resistance with her passion for public engagement and science communication, and is an author of the book *Science Communication: A Practical Guide for Scientists* (Wiley-Blackwell, 2012), and *The Microbes Fight Back: Antibiotic Resistance* (Royal Society of Chemistry, 2016). She currently leads an interdisciplinary research project focused on citizen science that is seeking solutions to the global problem of antibiotic resistance mechanisms (ARM). Laura also enjoys teaching medical sciences to future doctors at the Norwich Medical School.

Tom Husband

Tom Husband is a writer and educator who completed his Bachelor's degree in chemistry at Sheffield University. He has ten years' experience of teaching chemistry to students aged between eleven and eighteen, during which time he has studied for a Masters degree in Science Education at the University of Cambridge. His areas of pedagogical interest include metacognition, reflective writing, and collaborative reading. He is a regular contributor to various education publications, including hte Royal Society of Chemistry's Education in Chemistry, and also the American Chemical Society's ChemMatters. He has also published *The Chemistry of Human Nature* (Royal Society of Chemistry, 2016), a popular science book concerned with various aspects of biochemistry, including genetics, epigenetics, abiogenesis, and metabolism.

CONTENTS

ABBREVIATIONS

Å	Ångstrom $(1 \times 10^{-10}$ m$)$
ADP	adenosine diphosphate
AMP	adenosine monophosphate
AMPK	AMP-activated kinase
ATP	adenosine triphosphate
CAM	crassulacean acid metabolism
cAMP	cyclic adenosine monophosphate
DAG	diacylglycerides
dADP	deoxyadenosine diphosphate
dAMP	deoxyadenosine monophosphate
dATP	deoxyadenosiine triphosphate
DNA	deoxyribonucleic acid
FAD	flavin adenine dinucleotide
GI	glycaemic index
GTP	guanosine triphosphate
kJ/mole	kilojoules per mole
mRNA	messenger RNA
miRNA	mocroribonucleic acid
MRSA	methicillin-resistant *Staphylococcus aureus*
NADPH	nicotinamide adenine dinucleotide phosphate
nm	nanometre $(1 \times 10^{-9}$ m$)$
PCR	polymerase chain reaction
PEP	phosphenolpyravate
PKA	protein kinase A
PKU	phenylketonuria
PPK	phosphorylase kinase
PYG	glycogen phosphorylase
RNA	ribonucleic acid
rRNA	ribosomal RNA
tRNA	transfer RNA
μm	micrometre $(1 \times 10^{-6}$ m$)$
UTP	uridine triphosphate

1 CARBOHYDRATES: WHY LIFE IS SWEET

Life on Earth ranges from the very small, such as microscopic bacteria and protista, to the huge—think of blue whales and elephants (see Figure 1.1)! Yet all of these organisms are made from the same, microscopic, basic unit of life: the cell. The exact number, type, and arrangement of cells vary from one organism to another. They range from single-celled organisms, where one cell caries out all the functions of life, to highly specialized cells within a multicellular organism that perform particular functions, such as transmitting electrical impulses (nerve cells or neurons) or carrying out photosynthesis (palisade mesophyll cells). All cells are made from common building blocks, or biomolecules. The similarity in the make-up of cells also means that they carry out similar types of biochemical reactions, regardless of their cell type or function.

There are four types of fundamental biomolecules: carbohydrates, proteins, lipids, and nucleic acids; examples of each are shown in Figure 1.2. All of these biomolecules are examples of *organic* compounds, in which different chemical groups are attached to carbon-rich skeletal structures. While cells also contain inorganic compounds, among them calcium-, sodium-, and potassium-containing compounds, here we will be focusing on the biomolecules of life. This chapter describes the basic chemistry that underpins these biomolecules, and goes on to consider the amazing variety and flexibility of carbohydrates and their roles in living organisms.

Figure 1.1 From plants, birds, and mammals to fungi and bacteria—living organisms are all made up of the same fundamental types of biomolecules

(a)–(d): Anthony Short; (e): National Institute of Allergy and Infectious Diseases (NIAID)

Figure 1.2 The fundamental biomolecules: (a) a protein (in this case, a two amino acid dipeptide); (b) a nucleotide; (c) a carbohydrate (glucose); and (d) a lipid (cholesterol)

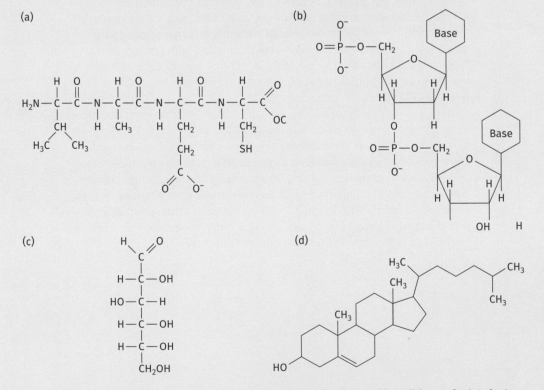

Adapted with permission from Crowe & Bradshaw, *Chemistry for the Biosciences*, third edition. Oxford: Oxford University Press, 2014. Reproduced with permission of the Licensor through PLSclear.

Life: it's elemental

Biochemistry takes us inside the fantastic inner world of living organisms and individual cells. But to understand biochemistry, we need a firm foundation in basic chemistry. By reviewing how atoms are held together, and how ions and molecules can influence each other, we will have the scientific knowledge we need to explain the biochemical basis of life. Life seems special, but it is important to remember that living things contain the same chemicals, and follow the same chemical rules, as inanimate objects. And yet, when these same fundamental building blocks are found in a living organism, they are part of a network of intricate, exquisitely regulated, and utterly complex chemical reactions that we call life!

The smallest 'unit' of every chemical element is the atom, which itself is made up of sub-atomic particles. These sub-atomic particles are protons (positively charged particles), electrons (negatively charged particles), and neutrons (particles with no charge). The protons and neutrons cluster together in a central nucleus surrounded by shells of electrons. Each element has a specific number of sub-atomic particles, giving each element its individual 'fingerprint' of structural and chemical properties. For example, a standard atom of carbon comprises six protons, six electrons, and six neutrons, while an atom of oxygen comprises eight protons, eight electrons, and eight neutrons. Some atoms can lose or gain electrons, leading to ions that are positively charged (cations) or negatively charged (anions).

For all chemical studies, including those of biological systems, it is important to understand some basic terms: the atomic number tells us the number of protons in the nucleus of an atom, whereas the mass number tells us the number of protons *plus* the number of neutrons. If atoms are uncharged, they contain the same number of electrons as protons, whereas charged atoms have increased or decreased numbers of electrons. Atoms with the same atomic number but different numbers of neutrons are known as isotopes; these have slightly different mass numbers even though they are the same chemical element.

The majority of elements found in nature exist as mixtures of isotopes, and the average mass of such mixtures of an element defines its standard atomic weight (also called the relative atomic mass). The atomic weights of elements are important in historical terms because these were the quantities that were measured by the earliest chemists.

The iconic image of the periodic table of elements is a beautiful summary of chemical information about all of the known chemical elements, and these details are essential for anyone studying chemical reactions, including biochemists. The periodic table orders the elements by their atomic number, as shown in Figure 1.3, with the elements divided into different rows (periods) and columns (groups). If we look across each row, from left to right, we see that each atom gains one extra proton and one extra electron.

The majority of chemical elements in the table were discovered in nature, although some are found in very small amounts. You might be surprised to find out that just over twenty different elements—about one-fifth of all elements found on Earth—are used to sustain the majority of terrestrial life. The commonest elements found in cells, from the most common

Figure 1.3 Periodic tables order different chemical elements according to their atomic number. They are separated into different periods (the rows) and groups (the columns). This is a standard periodic table, illustrating the name of the element, its chemical symbol, atomic number, and standard atomic weight.

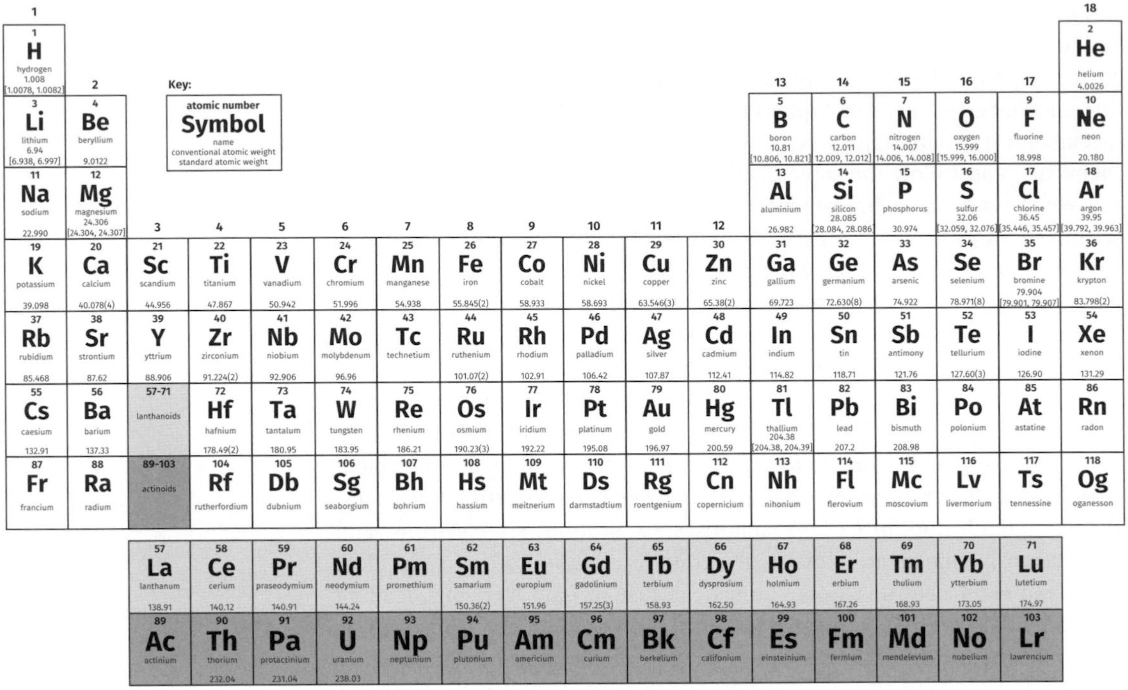

downwards, are hydrogen, oxygen, carbon, and nitrogen. The remaining elements typically make up just 2 per cent or so of all atoms in cells. From this exclusive list of elements, we can see that the majority of those that are useful to life have relatively low atomic numbers, but this is not the only factor that gives them a biological significance. Iodine, a key element used to sustain life, is extremely important for human health (see Chapter 5) and its atomic number is a relatively large 53.

The arrangement of elements in the periodic table helps us to understand relationships between their chemical properties. Within the periodic table, the elements on the left are often metals and those on the right are non-metals. Similarly, elements in the same column usually have similar chemical behaviours.

Each element in the periodic table has a defined number of protons and a matching number of electrons when it is uncharged. Electrons are located in what we call 'orbitals', which are grouped into 'shells'. Each shell can hold a precise number of electrons at set energy levels. Electrons fill the innermost shells first and each atom has maximum chemical stability when its shells are full. This is the case for atoms of all elements in the far-right column (group) of the periodic table (e.g. He, Ne, Ar, etc. in Figure 1.3). As a result of their full shells, atoms of these elements are chemically unreactive: we say that they are 'inert'. By contrast, atoms of elements whose outer shells are not full are *less* stable, making them *more* chemically reactive.

Put another way, the reactivity of any atom is dictated by the number of electrons in its outermost shells.

Covalent bonds

Clearly, not all atoms have enough electrons for all of their shells to be full. As an example, let's look at the position of oxygen in the periodic table (Figure 1.3). It is two electrons short of the full outer shell that makes neon (Ne) intrinsically stable. However, oxygen can achieve a full set of electrons in its outermost shell by sharing two additional electrons with other atoms. If oxygen shares an electron with two different hydrogens, all three atoms are able to complete their outer electron shells; the resulting molecule is a very stable chemical entity and is critical for life—you know it as water! When two atoms share a pair of electrons a covalent bond is formed.

Ionic bonds

Another important type of bond is the ionic bond, which occurs when one atom completely gives up an electron to another atom. When two atoms pair up this way, the resulting compound is called a 'salt'; perhaps the best-known example in biological systems is sodium chloride—known as 'table salt' to you and me! The periodic table (Figure 1.3) shows us that sodium has one electron in its outermost shell, whereas chlorine is one electron short. Using this type of analysis, it becomes clear that if sodium donates its outermost electron to chlorine, both atoms have full outermost shells. This compound of sodium chloride is very stable at a chemical level.

Other ways of bonding

A number of other types of chemical bonding play important roles in both biological and non biological systems. One of these is hydrogen bonding, which results from an electrostatic attraction between a proton in one molecule and an electronegative atom in another (such as nitrogen, oxygen, or fluorine). These interactions take place over short distances—the length of a typical hydrogen bond is about 0.2 nanometres (nm), or 2×10^{-10} m—and are influenced by other atoms that are nearby, as you can see in Figure 1.4. Compared to ionic and covalent bonds, each hydrogen bond is relatively weak. But hydrogen bonds occur very frequently, and the combined effect of many weak bonds can have a big impact: the relatively high boiling and melting points of water are the result of the hydrogen bonds holding the molecules together.

Hydrogen bonds play a major role in stabilizing the structures of molecules, which are often important for their function. For example, hydrogen bonds play an important role in the three-dimensional structures of proteins, including the active sites of enzymes, and the DNA double helix itself, as you will see in Chapters 2 and 3.

Carbon is a fascinating atom, which is absolutely essential for all life on Earth. With a total of six electrons, carbon has four electrons in its outermost shell. In chemical terms, this provides a dilemma and an opportunity: to achieve a full outermost electron shell should it donate, accept, or share

Figure 1.4 (a) Diagram to show the structure of hydrogen bonds, with the letter δ indicating partial positive and negative charges on the H and O atoms. (b) Hydrogen bonds are important in the formation of ice crystals by water molecules in winter—beautiful but potentially damaging to living organisms.

(a)

(b)

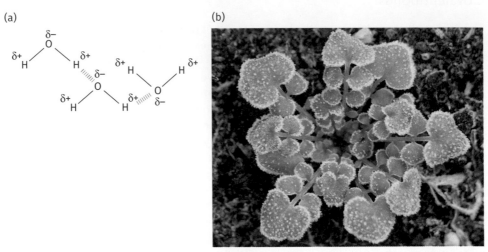

(b): Anthony Short.

four electrons? Carbon has opportunities to form covalent bonds with four other atoms, which can be other carbons or a large range of other atoms, including many of the nineteen other elements that are required for life. These bonding patterns allow the large and intricate molecules we find in cells to form and enable the chemical reactions that are fundamental for life.

Many important biological molecules are **polymers**. A polymer is a large macromolecule made of lots of identical or similar smaller molecules called **monomers** (see Table 1.1).

Table 1.1 Polymers and monomers of common biomolecules

Proteins	Polymer = polypeptide	Monomer = amino acid
Carbohydrates	Polymer = polysaccharide	Monomer = monosaccharide
Nucleic acids	Polymer = polynucleotide	Monomer = nucleotide

Two monomers join to form a **dimer**, several monomers join to form an **oligomer**, and lots of monomers join to form a **polymer**.

So we have:

Monosaccharide, disaccharide, oligosaccharide, polysaccharide

Nucleotide, dinucleotide, oligonucleotide, polynucleotide

Amino acid, dipeptide, oligopeptide, polypeptide.

For these molecules, the monomers join together by a reaction called **condensation**, a reaction where a molecule of water (H_2O) is lost. A covalent bond forms between the monomers (see later in this chapter, and Chapters 2 and 3). Polymers are broken down into monomers in a **hydrolysis reaction**, with the *addition* of a molecule of water.

Figure 1.5 The structure of a triacylglycerol

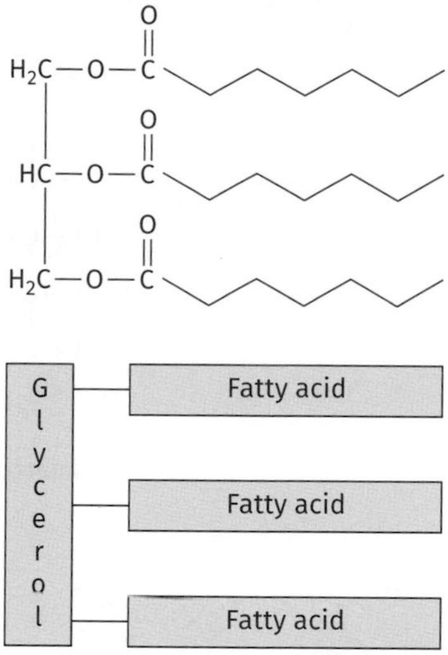

Adapted with permission from Crowe & Bradshaw, *Chemistry for the Biosciences*, third edition. Oxford: Oxford University Press, 2014. Reproduced with permission of the Licensor through PLSclear.

Lipids are *not* polymers—they are not made of lots of monomers. Triglycerides are made of one glycerol and three fatty acids, as shown in Figure 1.5. But they are big, so they are macromolecules.

One of the groups of carbon-rich molecules found most widely in living organisms are the carbohydrates. Armed with the basic biochemistry knowledge introduced above, read on to discover more about these amazing molecules.

The structure of carbohydrates

Perhaps carbohydrates deserve the title of the building blocks of life more than any other type of biomolecule. As the name suggests, these molecules are simply hydrated carbon atoms, which means they are made from just three elements: carbon, hydrogen, and oxygen, with the general molecular formula $(CH_2O)_n$ for the simplest carbohydrates. Also known as saccharides, from the Greek for sugar, these simple building blocks include the mono-saccharides $((CH_2O)_n$, where n is 3 or more) such as glucose and galactose, and also the disaccharides that include sucrose, lactose, and maltose (see Figure 1.6). Until recently, carbohydrates were understood to be important fuels and structural components of all cells, but they were thought to have little influence over key biochemical reactions. However, we now know that carbohydrates support the biochemical architecture of cells and provide many of the fundamental characteristics that define the way each type of cell works. They are even essential components of some protein molecules, allowing the protein to function correctly, and play key roles in cell-to-cell communication. The surge of interest in these molecules has led to the development of new fields of biochemistry, known as glycobiology and glycomics.

Figure 1.6 The structures of some carbohydrates—the monosaccharides glucose and galactose, and the dissacharides sucrose and lactose. Notice how lactose is, in fact, a dimer of glucose and galactose.

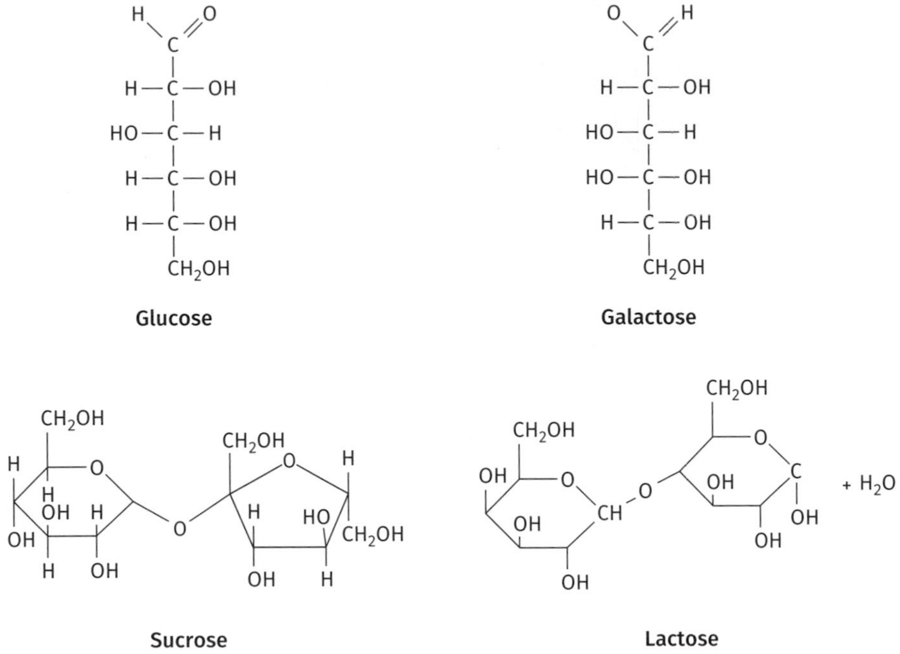

Glucose

Galactose

Sucrose

Lactose

Adapted with permission from Crowe & Bradshaw, *Chemistry for the Biosciences*, third edition. Oxford: Oxford University Press, 2014. Reproduced with permission of the Licensor through PLSclear.

The massive chemical and structural diversity of carbohydrates is key to their importance for cells.

The names of these many different, closely related compounds provide clues to their chemical structures. It is not essential to know the chemical structure associated with all the names, but it is useful to know a few key details. Monosaccharides are usually classified by the number of carbon atoms they contain: triose sugars contain three carbon molecules; tetrose, pentose, and hexose sugars contain four, five, and six carbons, respectively.

The various sugars are important for different types of biochemical reactions but the most important monosaccharide for most cell types is glucose, a hexose. During photosynthesis, plants and some bacteria use energy from light along with clever biochemical reactions to convert carbon dioxide in the atmosphere into glucose molecules. In turn, glucose molecules are broken down to carbon dioxide and water during cellular respiration in most cells.

Monosaccharides are often described as simple sugars, but looks can be deceptive: they are more complex than they might appear at first glance. Since different chemical groups are linked to the carbons, they are **chiral** or asymmetric molecules, although most naturally occurring monosaccharides exist as one type, D-isomers (see Figure 1.7). But what do we mean by the term 'isomer'?

Figure 1.7 Comparison of the chemical structures of D-glucose, D-fructose, and D-galactose. The chemical structures are drawn to represent the three-dimensional structures of carbohydrates in two dimensions. Since carbohydrates contain chiral carbons, they exist in different isomers. To determine the type of isomer of a specific carbohydrate, they are drawn with the carbonyl group (carbon double-bonded to an oxygen) at the top, as shown here. Reading away from the carbonyl group, the OH group on the penultimate carbon determines the isomer: if it is pointing left, then it is 'L'; if it is pointing right, then it is 'D', as shown here.

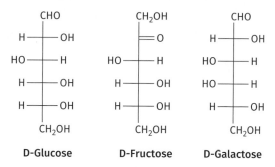

D-Glucose   **D-Fructose**   **D-Galactose**

To explore more, let's look at the monosaccharides glucose, galactose, and fructose. These sugars share an identical chemical formula, $C_6H_{12}O_6$, yet they are not identical! Their differences lie in the way the individual oxygens and hydrogens attach to the carbon backbone. What does this mean in practice? Let's look first at glucose and galactose. Even if you draw these sugars using standard chemical projections, it is hard to see the difference between them. But when you look closely, you can see that the hydroxyl (–OH) group that is bonded to the fourth carbon from the top of each structure has switched to the opposite side of the carbon backbone. These molecules are known as isomers. It is a subtle chemical difference that can be hard to spot, but it causes an extremely important change to the biochemical properties of these molecules, including their melting points and solubility in water (Figures 1.7 and Table 1.2).

The monosaccharide fructose also shares the same chemical formula. It is found in many plants and fruits (leading to the name fructose, 'fruit sugar') and honey, and it is much more soluble than the other, similar sugars. Its ability to dissolve in water means that it is quickly absorbed through the human intestine and into the blood stream during digestion. Although fructose has the same chemical formula as glucose and galactose, it has a structural difference that is easier to spot (Figure 1.7). The position of the carbon–oxygen double bond is different—in fructose it involves the second carbon in the chain, but in glucose and galactose it involves the first carbon in the chain.

The three monosaccharides shown in Figure 1.7 exist in an equilibrium of different chemical structures. While they can adopt the linear form shown in Figure 1.4, they are also able to form a six-atom ring (or in the case of fructose, a five-atom ring), when the alcohol group (–OH) at one end of the monosaccharide bonds to the carbonyl group (–C=O) at or near the other end, forming a cyclic compound (Table 1.2). In fact, molecules with five or more carbons are more likely to adopt the ring structures in aqueous solutions and, therefore, inside cells.

Table 1.2 The diversity of chemical structures and biochemical properties of simple saccharides

Saccharide name	Type of saccharide unit	Chemical structure	Solubility in water (g/L, at 25 °C)	Melting point (°C)
Glucose	Monosaccharide		909	146–150
Galactose	Monosaccharide		650	163–165
Fructose	Monosaccharide		4000	112–113
Maltose	Disaccharide of two glucose molecules		1080	102–103
Lactose	Disaccharide of glucose and galactose		195	202
Sucrose	Disaccharide of glucose and fructose		2000	None (decomposes at 186°C)

Sweet sensations

Many animals—including humans—like sweet things. This sensory effect has evolved in response to the qualities of glucose, fructose, and galactose, as well as the disaccharides that form when they bond together. They all taste sweet. Some scientists think that taste emerged to help make sense of the food consumed for survival. A bitter or sour taste is

Figure 1.8 This green basilisk lizard is prepared to risk being seen to eat a sweet banana with its rich supply of sugar. In this case, it won't be dispersing seeds—people left the banana to lure the lizard into view, and the fruit has no seeds!

Anthony Short.

a good indication that a plant should be avoided as a food source as there is a strong likelihood that it contains natural products that may be poisonous or harmful. On the other hand, it is thought that many animals are attracted to sugars because of the need to identify readily available sources of metabolic fuel, especially glucose. Plants that need insects to pollinate them often lure the insects to the flowers by producing a sweet reward in the form of nectar. Similarly, plants which produce sweet fruits attract a whole range of animals that all rely on glucose as an energy source. By attracting animals ranging from mammals to insects to consume their fruit, with the promise of a sweet taste and a sugar rush, plants use animals to spread their seeds and help them to germinate (see Figure 1.8).

We can see the significance of these carbohydrates for the evolution of multicellular organisms by using humans as an example. We are born liking sweet-tasting substances, and human milk is very sweet. We can taste 'sweetness' using microscopic bumps—taste buds—found on the mucous membrane of our tongues. Our taste buds are made up of a bunch of papillae that contain sensory nerve cells. The chemical substances that provide sweet taste (such as glucose or fructose) are freed as we chew food and add saliva. At the top of each taste bud is a small indentation filled with fluid that dissolves saccharides, and ensures that as many sensory nerve cells as possible embedded within our taste buds are stimulated by the sugar molecules. The cellular membranes of each sensory nerve contain specific receptor molecules, a dimer of two different protein receptors

Figure 1.9 Hummingbirds live on nectar and they can taste sugars because natural selection refashioned a savoury receptor that detects amino acids into a unique sugar receptor

(a)   (b)

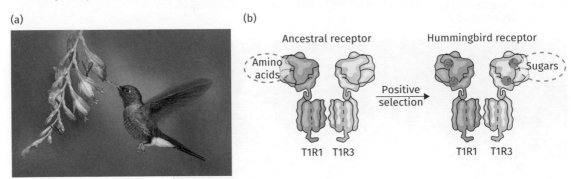

(a): Ondrej Prosicky/ Shutterstock.com; (b): V. ALTOUNIAN/SCIENCE; (b) Adapted with permission from Peihua Jiang, G. and Beauchamp, G. K., 'Sensing nectar's sweetness', *Science*, **345(6199)**, 878–9, 22 August 2014. Copyright © 2014, American Association for the Advancement of Science.

referred to as TR1s, which contain different versions of the R molecule (see Figure 1.9). Saccharides bind to these receptors and, in so doing, activate these nerve cells. The receptor cells then activate further nerve cells that carry information to the brain, which perceives a sweet taste.

Perhaps surprisingly, the sweetness receptors can also bind aspartame. This is not a sugar, it is a dipeptide; yet it also stimulates the nerve cell to detect sweetness. This is why aspartame is often used as an 'artificial' sweetener, and can be useful if you want to lose weight: it provides a sweet taste with far less energy than the saccharides. The importance of saccharides in cellular metabolism will be discussed in more detail in Chapter 4.

Interestingly, domestic and wild species of cats are exceptional because they show no interest in sweet foods: as obligate carnivores (that is, meat-eaters) they have no evolutionary need to detect sweet sugars in plants. When scientists studied the biochemical structure and function of the T1R2-T1R3 receptor molecules in cats they discovered that the gene that coded for the T1R2 protein has several mutations which means that the receptor dimer is no longer fully functional. When scientists tested other obligate carnivores, such as spotted hyenas and sea lions, as well as many species of carnivorous birds that eat insects or other animals, they found that these species had also lost functional T1R2 receptor proteins, and with it their 'sweet tooth'. This evidence suggests that when animals no longer rely on plant material as a source of energy, a sweet tooth is no longer required and so the evolutionary pressure to retain the receptors is lost.

There are, however, many species of birds that rely on flowers and fruit as a food source. Hummingbirds are a well-known example, relying on nectar as food. Nectar is a source of almost pure sugar. However, scientists were surprised to discover that they also have dysfunctional sweetness receptor molecules. Hummingbirds have evolved a way to compensate for this by repurposing another, closely related, receptor (T1R1-T1R3) to detect sugars instead, generating their own ecological niche that is closed to other non-sweet-detecting birds. This is summarized in Figure 1.9.

Lacking lactose

Animals produce milk to feed their young, and the milk of many species contains lactose. This lactose is important because it can be readily metabolized to release glucose. It doesn't take a great leap of imagination to see why there is an advantage for having a sweet-tasting milk, as this will be more likely to be consumed by infants. Lactose is metabolized to individual monomers of glucose and galactose by the enzyme lactase, which is found in the small intestine. Humans cannot absorb lactose through the small intestinal walls, so if it fails to be metabolized by lactase it passes through the small intestine into the colon, and this is where problems can begin. As lactose enters the colon, it encounters a range of bacteria that use fermentation pathways to break it down, releasing a mixture of hydrogen, carbon dioxide, and methane gases that can cause abdominal discomfort and bloating. Any sugar that is unabsorbed, as well as some of the other fermentation products, increases the osmotic pressure within the colon. In turn, this causes water to move into the bowel and the result is diarrhoea.

The genetic instruction for producing the lactase enzyme is provided by the *LCT* gene. This gene is turned on (and off) by a protein called MCM6. This regulation ensures that lactase is produced in infants, who need the enzyme to metabolize the lactose found in breast milk. Several thousand years ago, some humans developed a mutation in the *MCM6* gene that keeps the *LCT* gene turned on even after breast feeding is stopped, allowing them to metabolize lactose even in adulthood. This population is described as having lactase persistence. Genetic analysis on a global scale shows that this lactase persistence has evolved independently several times in several different places. Other populations (without this mutation) stop producing lactase when breast feeding stops and, as members of this population age, the *LCT* gene is turned off, lactase stops being produced, and they lose their ability to metabolize lactose. These populations are described as having lactose intolerance and the level varies globally. Less than 10 per cent of the population in Northern Europe is lactose intolerant, while more than 90 per cent of the adult population in parts of Asia and Africa cannot digest milk. (For more on epigenetics, see *Genomics* in this series.)

Although milk contains up to 5 per cent lactose, significant quantities of lactose are removed when it is turned into butter. Yoghurt and cheese are also dairy products, but both are the result of fermentation reactions using bacteria (such as lactobacilli) that remove lactose using their own lactase enzymes. This is why many lactose-intolerant individuals can tolerate butter, cheese, and yoghurt, but are unable to tolerate milk.

Glucose polysaccharides

Nature has exploited the subtle differences that exist between monosaccharides. Glucose is one of the most prevalent molecules in nature, and is used as a source of energy in cells (including brain cells), and it is one of the building blocks of common disaccharides such as sucrose and lactose. (We will learn more about glucose in Chapter 4.) It is also the basic building block of some of the more complex carbohydrates (polysaccharides) that are used as energy stores, such as starch and glycogen, and for cellulose, a molecule that provides physical strength and rigidity to plants.

The different functional properties of starch and cellulose depend on the bonds that form between the individual glucose molecules. Starch is created from glucose molecules that polymerize: they are joined together by α bonds as a result of a condensation reaction (dehydration) that takes place between individual glucose monomers. These α bonds can be easily broken or hydrolysed. This ease of bond-breaking is also shared by glycogen, the polysaccharide used by animals to store spare glucose molecules as a future energy source.

On the other hand, the glucose monomers within cellulose and chitin (the main constituent of shells or exoskeletons of arthropods and the cell wall of fungi) are held together by β bonds, which are much more stable and resistant to hydrolysis (Figure 1.10).

Currently, there is much debate about 'good' and 'bad' carbohydrates in our diet. The glycaemic index (GI) is used to rank carbohydrate in foods according to how they affect blood glucose levels (See Scientific approach 1.1). Carbohydrates with a low GI value are digested and metabolized relatively slowly, which causes blood glucose levels to rise more slowly too. This results in the body producing lower overall levels of insulin compared to foods with a high GI.

Figure 1.10 Glucose molecules can link to each other using different types of bonds. Different types of complex polymers of glucose are formed using different bonds, with α bonds in starch (highlighted in red) and β bonds in cellulose (highlighted in blue). The β bonds are more resistant to hydrolysis, which is helpful for stable structures such as cell walls.

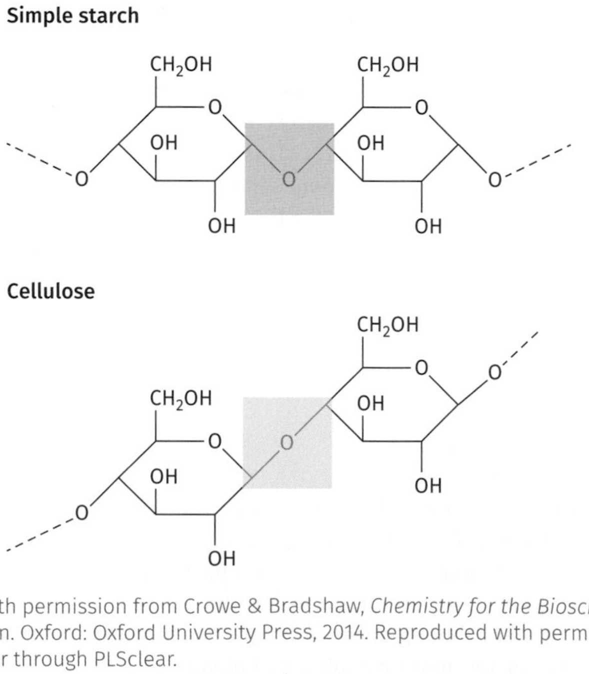

Simple starch

Cellulose

Insulin is a peptide hormone released from beta cells in the pancreas. One of its key roles is to enable tissues like liver, fat, and skeletal muscle to take in the glucose circulating in the blood stream. These tissues can convert glucose into glycogen and fat-storage molecules. Western diets often contain foods with a high GI, which lead to high levels of insulin production that flood the blood stream. However, our bodies are unable to cope with long-term high levels of glucose. Initially, the pancreas produces extra insulin, but over time it can't make enough insulin to maintain normal blood glucose levels. This is called insulin resistance and is the reason that foods with a high GI have been linked to an increased risk for health problems, including type 2 diabetes, heart disease, obesity, age-related macular degeneration, infertility, and colorectal cancer. Maintaining a diet that depends on foods with a low GI can help to control diabetes and improve weight loss.

Scientific approach 1.1
Measuring glycaemic index

It is important to measure the glycaemic index (GI) value of foods accurately. It is impossible to correctly assess the composition of food by using guess work or relying on taste, smell, size, and shape. The International Standard method for accurately calculating the GI value of food is to feed a minimum of ten healthy people a portion of the **TEST** food containing 50 grams of digestible carbohydrate. The blood glucose level of the healthy participants is measured for the next two hours (Figure A). After a sufficient time has passed, the same participants are asked to consume an equal-carbohydrate portion of glucose, the **REFERENCE** food, after which their two-hour blood

Figure A Graphs indicate typical results for blood sugar levels after eating foods with high (red line) and low (blue line) Glycaemic Index. The amount of carbohydrate in the reference and the test food must be the same when the GI is measured.

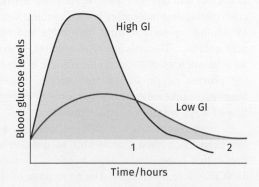

glucose response is measured once again. A GI value for the test food is then calculated by comparing the results of its glucose levels to the results of the glucose reference food. The final GI value for the test food is the average GI value for the ten people.

Foods that have a high GI score contain carbohydrate that will be rapidly digested, which produces a large rapid rise and fall in the level of blood glucose. In contrast, foods with a low GI score contain slowly digested carbohydrate, which produces a gradual, relatively low rise in the level of blood glucose.

 Pause for thought

Starchy foods like rice and pasta have a high GI compared to beans, fruits, and vegetables, which contain more cellulose and fibre. From your knowledge and understanding of the polysaccharides contained in these food substances, provide a biochemical explanation for this difference.

Celebrating cellulose

Plants are not just an excellent source of food for people and other animals. They also produce other natural products that can be useful. Cellulose is a linear polymer created from glucose molecules; it is a key component of plant cell walls and is one of the most abundant polymers on Earth. At first glance, this natural product might appear to be a rather stiff, boring, odourless, and tasteless molecule. But, in fact, it is these properties that offer a wealth of possibilities that we can exploit to provide materials that support modern human life.

The books we read—including this one (unless you're reading this as an ebook!)—are printed on paper, and a key component of paper is cellulose. The cellulose found in wood pulp has chain lengths of 300–1700 glucose molecules and does not dissolve in water. Breaking it down into smaller glucose-containing molecules is challenging. Cellulose pulp is depolymerized by adding aqueous sodium hydroxide and carbon disulfide to the liquid pulp solution. The polymer that remains, viscose, can then be treated with an acid, such as sulfuric acid, before it is spun into rayon fibres. However, the process of producing this natural fibre has environmental consequences: it takes significant amounts of energy and results in the production of large volumes of waste products.

In more recent times, a different solvent, 4-methylmorpholine 4-oxide, has been discovered that is able to dissolve cellulose, producing a solution called dope. When dope is added to water it precipitates to produce TENCEL®, a branded fibre. The advantage of this process is that more than

99 per cent of the solvent is recovered and reused, making this a much more environmentally friendly process.

Viscose can also be treated with sulfuric acid and sodium sulfate to create cellophane, a flexible transparent sheet of glucose polymer that is waterproof and used as an alternative to clingfilm to protect food. It is also the basis of the famous sticky substance found in brands such as Sellotape.

Another well-known type of cellulose fibre is cotton, a soft fibre that forms a cushion around the seeds of the cotton plant (see Figure 1.11). It is an incredibly pure source of natural cellulose fibre that has been used for millennia in different human societies as a lightweight fibre that can be spun into cloth. Cotton production accounts for almost 1.5 per cent of the world's arable land that is not being used to grow food, supporting the 10 million bales of cotton that are used annually. As a crop, it is well suited to the seasonally dry tropics and subtropics in the Northern and Southern hemispheres, but its cultivation in areas with low rainfall has led to use of irrigation practices that have a damaging effect on limited water resources. The production of cotton also relies on chemicals such as herbicides, fertilizers, and insecticides, which are used to improve yields and destroy common pests such as the boll weevil and the cotton boll worm. Genetic modification has generated a cotton plant, GM Cotton, containing a bacterial gene that allows the cotton plant to produce a protein that is toxic to these pests and reduces the reliance on harmful pesticides. Organic cotton, on the other hand, is GM-free and is certified to be grown without the use of any synthetic agricultural chemicals.

Figure 1.11 Cotton is a soft fibre that forms a cushion around the seeds of the cotton plant

Natalia Kuzmina/Shutterstock.com

Ruminating about ruminating

Plants are not just an excellent source of sugar-rich nutritional material such as nectar and fruits. Substantial amounts of vegetation are used as the main food source for many animals. These plants contain significant amounts of polysaccharides, such as cellulose. As we have already seen, cellulose is a linear polymer created from glucose molecules, and it is a key component of plant cell walls. But cellulose is a tough molecule with bonds that are hard to digest—so how do animals manage to digest it?

The majority of mammals do not produce enzymes that can break down cellulose, yet many of them live by feeding on grass or other leaves. One group of grass-eating mammals, the ruminants, include many familiar animals—e.g. cows, sheep, and goats. The guts of these animals have adapted to form an excellent environment for a range of microbes that can ferment cellulose. This fermentation occurs at the start of the animal's digestion system, creating cud, which is regurgitated back and chewed again, breaking down the plant matter even further. These animals literally 'chew the cud'. This process of re-chewing and the further digestion it stimulates is known as rumination (from the Latin *ruminare*, which means 'to chew over again'). Rumination also helps to physically break down the plant cells, releasing the contents so they can be acted on by standard gut enzymes.

The microbiome of ruminants is complex and contains bacteria, protozoa, and yeast. Most of these microbial species do not need oxygen to survive. Instead, they anaerobically ferment plant material, such as cellulose, producing sugar molecules along with other organic molecules such as acetate, lactate, and methane. It is estimated that there are up to one thousand billion, or 10^{12} (1,000,000,000,000) organisms *per millilitre* of ruminant fluid.

The bacteria that ferment the plant material also consume some of the energy and molecules that they release from the plant material, which is not beneficial to the mammal. But these ruminants have another adaptation that is helpful. As the bacteria pass through the gut beyond the rumen, they are attacked by enzymes such as lysozyme and ribonuclease. This allows the ruminant to digest the bacteria, releasing their energy and molecules, and therefore helping the mammals to thrive. Not all plant eaters are ruminants, however (see Case study 1.1).

Case study 1.1
The panda

Giant pandas are some of the most easily identified and loved mammals on Earth, and it is well known that bamboo forms almost all of their diet (Figure A). Giant pandas that were on Earth around 7 million years ago were omnivores and only started to incorporate bamboo into their diet relatively recently. They began to rely on bamboo as their sole source of food about

2 million years ago and evolved strong jaw muscles and a 'pseudo thumb' as a result of selection pressure: both of these adaptations help them tackle this tough, cellulose-laden plant. But unlike other herbivores such as sheep and cows, their digestive system failed to adapt, and the intestines of modern pandas are almost the same as their ancient panda relatives.

A study undertaken by Chinese scientists in 2015 collected the faeces from forty-five different pandas of variable ages for a whole year. By sequencing the pieces of RNA found in the pandas' poo, they discovered the makeup of microbes that were found in panda guts—they unravelled the panda's microbiome. What is surprising is that their intestines are relatively short, like most carnivores, and it takes less than half a day for food to travel from mouth to anus. What was even more surprising is that, unlike other herbivores such as horses and sheep, panda guts contain *Escherichia*, *Shigella*, and *Streptococcus*, bacteria normally found in the guts of carnivores. Unlike the microbes found in the guts of ruminants, these bacteria are not known for their ability to break down cellulose to release these energy-rich glucose molecules.

This result led to consternation within the scientific community, as it seemed to make no biological sense. Panda DNA contains no genetic material that codes for enzymes to break down cellulose and their guts hadn't acquired the microbes that would do this job for them. How had pandas been surviving for millions of years on a diet of cellulose-rich bamboo?

A scientific study published in 2018 provided some clues. Pandas and their associated microbiome contain no enzymes that can digest cellulose, but they *do* contain an abundance of genes that code for enzymes that degrade

Figure A Giant pandas live almost exclusively on bamboo

leungchopan/Shutterstock.com.

starch and hemicellulose. In fact, Panda microbes contain more of these genes than their counterparts found in other herbivores and omnivores. This implies that pandas do not rely on cellulose to obtain energy, but they seem to be effective digesters of starch and hemicellulose. Hemicellulose is a much smaller molecule than cellulose and contains many different sugar monomers; it also has a more varied structure than cellulose, with little strength, and it is easy to digest by hemicellulose enzymes.

Scientists had observed that, given a choice, giant pandas prefer to feed on the tender shoots and leaves of young, year-old bamboo plants rather than on older, larger plants. These bamboo elements are high in hemicellulose and contain starch that can be broken down by the giant panda's metabolism. The gut microbiome of pandas has not adapted to degrade cellulose, but it has adapted to make best use of the starch and hemicellulose found in young bamboo shoots and leaves—and panda behaviour has evolved to supply the microbes with the young bamboo they can best digest.

In this chapter, we have seen that life on Earth is dependent on a wide range of chemical elements. Using these small chemical components, all cells construct key molecules to form common building blocks, which are used to generate the biomolecules that are essential for life. Each type of molecule has a variety of chemical groups attached to different skeletal structures and these influence the plethora of biochemical reactions that occur in all cells, regardless of their cell type. This chapter has highlighted the amazing variety and flexibility of carbohydrates, focusing on their roles in living organisms. In the following chapters, we look at other groups of chemicals and the complex biochemistry that is behind life itself.

 Chapter summary

- All biological cells are made up of common building blocks, or biomolecules, which carry out similar types of biochemical reactions. A common feature of these molecules is that they are polymers constructed from monomers of certain types of biomolecules.
- There are four types of these fundamental biomolecules: carbohydrates, proteins, lipids, and nucleic acids.
- All chemical elements are composed of atoms, which contain protons, electrons, and neutrons. The numbers of each of these sub-atomic particles are specific to each element, leading to distinct structural and chemical properties.
- Electrons are arranged in configurations termed 'shells' that can each hold a specific number of electrons at set energy levels. Each atom has maximum chemical stability when its shells are full, meaning that its reactivity is dictated by the number of electrons in its outermost shell.

- If two atoms have a complementary number of electrons, they can *share* two electrons to form a covalent bond. In ionic bonds, one atom gives up one (or more) electrons to another atom.

- Carbohydrates are the most abundant biomolecules on Earth, and are characterized by great chemical and structural diversity. They are classified by size as monosaccharides, disaccharides, oligosaccharides, or polysaccharides.

- Carbohydrates are chiral molecules, although most naturally occurring monosaccharides exist as one type, D-isomers. Most monosaccharides with five or more carbons are more stable as cyclic molecules.

- The polymerization of monosaccharides to form di- and polysaccharides happens via a condensation reaction (dehydration).

- Carbohydrates have important biological roles in all living organisms. They are found in plant cell walls, connective tissues of animals, and exoskeletons, and they are essential to the function of some proteins and cell-to-cell communication. Carbohydrates are also important fuels and energy stores, underpinning energy-yielding pathways in non-photosynthetic organisms.

Further reading

Beauchamp, G. (2016) 'Why do we like sweet taste: A bitter tale?', *Physiology & Behavior*, **164** (Pt B), 432–7.
doi: 10.1016/j.physbeh.2016. 05.007
https://www.ncbi.nlm.nih.gov/pmc/articles/PMC5003684.

Prescott, L. M., Harley, J. P., and Klein, D. A. (2005) *Microbiology*, sixth edition. McGraw-Hill, New York.

Wilkinson, A. (2015) 'Nature news: Panda guts not suited to digesting bamboo' (**doi:10.1038/nature.2015.17582**).

Zhang, W., Liu, W., Hou, R. et al. (2018) 'Age-associated microbiome shows the giant panda lives on hemicellulose and not cellulose', *The ISME Journal*, **12**, 1319–28. **https://doi.org/10.1038/s41396-018-0051-y.**

http://www.periodicvideos.com: an online resource, including videos that describe interesting chemical properties of all 118 elements.

https://www.glycopedia.eu: a useful online resource, providing background information about a wide range of carbohydrates.

Discussion questions

1.1 Some elements of lower atomic weight are not useful for cellular life, including those from Group VIII. What are the chemical properties of Group VIII atoms that explain this?

1.2 Covalent and ionic bonds are critical for the correct conformation of all biomolecules. Describe three other types of interactions that occur in and between biomolecules in living cells.

1.3 Discuss ethical arguments for and against the use of organic cotton and GM cotton. Having considered both, suggest which you think should be used where possible and explain your decision.

1.4 Would you class Rayon, Viscose, and TENCEL® as natural or manmade fibres? Explain your decisions and discuss how these products are perceived among wider society.

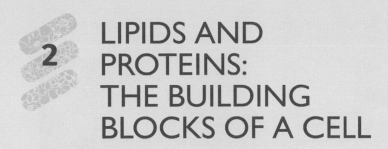

2 LIPIDS AND PROTEINS: THE BUILDING BLOCKS OF A CELL

It is often stated that cells are the basic unit of life, but cells require essential building blocks if they are to undertake the huge range of biochemical reactions needed for life, and to provide the structural scaffolding required to keep the cells together and the reactions apart. All living cells contain a large number of complex molecules that need to be synthesized, transformed, metabolized, or taken up from the environment and transported across the cell membrane. Proteins and lipids are two of the key components of cellular life in every environment (see Figure 2.1).

You may have heard the phrase 'essential fatty acid'. This phrase stems from the fact that these molecules are vital for cellular life. For us humans, fats contain components that we cannot synthesize ourselves; we need to absorb them from our diet. The same is true for some amino acids, the starting points for all **proteins**, a huge family of molecules that come in many shapes and sizes. Proteins undertake key roles, including the **enzymes** that catalyse biochemical reactions, sensor molecules that pick up and pass on environmental signals, and structural molecules that enable the cell to maintain its shape and therefore its function. It is essential for us to understand how lipids and proteins are synthesized, and how structure influences their function, if we are to understand how cells act as the basic unit of life.

Figure 2.1 The lipids and proteins making up the membranes and enzymes of cells have to be able to function in all extremes of nature

Anthony Short.

Lipids and fatty acids: keeping everything in working order

Lipids are a diverse group of naturally occurring substances that perform many different important roles in living organisms. Different chemical groups attached to different lipids mean they have varying levels of solubility in water, but the one thing they all have in common is their ability to dissolve in a non-polar solvent instead of water. We are most familiar with lipids being essential components of cell membranes, but they are also used to store energy. Lipids are also able to provide a waterproof coating to feathers, skin, and fur, and they are the basic building blocks of a raft of molecules we rely on for our survival; these include chemical signals such as hormones, vitamins, and pigments.

Lipids can be classified in different ways, but they include:

1. fatty acids;
2. triglycerides;
3. isoprenoids;
4. phospholipids;
5. sphingolipids; and
6. wax esters.

These terms are quite complicated, so in this chapter we will explain what each one means. We will also look at their structures and review the biochemical functions of examples of each.

Fatty acids

Fatty acids are long chains of hydrocarbons that are typically between twelve and twenty carbon atoms long. A complicated naming system describes the different types of molecules, but we will familiarize ourselves with just the basic, most useful details (Figure 2.2).

Each fatty acid contains a methyl group (–CH$_3$) at one end of the molecule and a carboxyl group (–COOH) at the other end, with a hydrocarbon chain in between. Figure 2.2 outlines the structures of different types of naturally occurring fatty acids. Although the arrangement of atoms in the fatty acids of different organisms varies widely, most naturally occurring fatty acids tend to have an even number of carbon atoms. If these chains contain nothing but C–C single bonds, they are described as saturated fatty acids, but if they contain one or more C=C double bonds they are said to be unsaturated fatty acids (Figure 2.2(b)). A fatty acid with a single C=C bond is described as a monounsaturated fatty acid, whereas those with two or more C=C bonds are referred to as polyunsaturated fatty acid.

Figure 2.2 Commonly used nomenclature for fatty acids. (a) The chemical structure is shown in relation to Greek letters. The methyl (CH$_3$–) end is defined as the omega (ω) end, while the carbon atom next to the carboxyl (–COOH) group is called the alpha (α) carbon, and the subsequent one the beta (β) carbon. The letter *n* refers to the remaining number of carbons. (b) Chemical structure of several unbranched fatty acids containing eighteen carbon atoms, indicating whether they contain zero, one, or more double bonds (shown with *cis* configuration).

(a)

$$CH_3 — (CH_2)_n — CH_2 — CH_2 — COOH$$

ω   β   α

(b)

Common name	Chemical structure	Type of fatty acid
Stearic acid	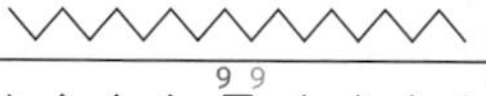	Saturated
Oleic acid	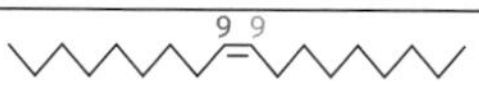 9 9	Monounsaturated
Linoleic acid	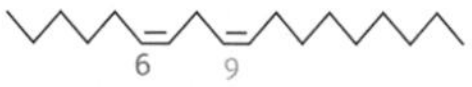 6 9	Polyunsaturated
α-Linolenic acid	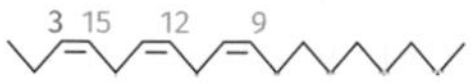 3 15 12 9	

Figure 2.3 *cis- and trans-*isomers of a fatty acid

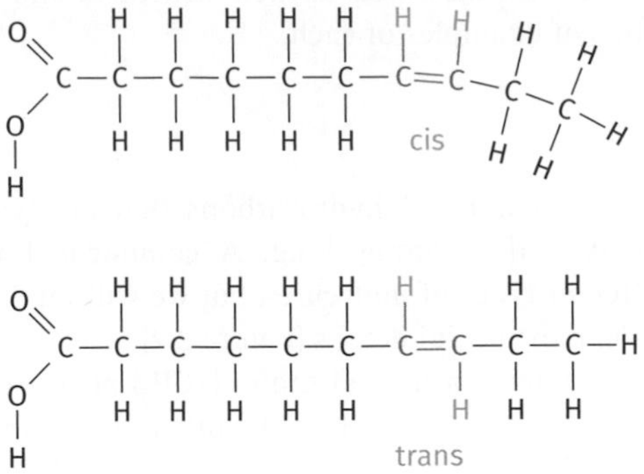

Double bonds come in two distinct forms: they can either have their attached chemical groups on the same side of the double bond, or on opposite sides. Those fatty acids with attached groups on the same side of a double bond are called *cis* fatty acids (they are the *cis*-isomer); those with groups on opposite sides of the double bond are *trans* fatty acids (they are the *trans*-isomer). This difference is illustrated in Figure 2.3. Naturally occurring unsaturated fatty acids almost always have a *cis* conformation. This form of bond causes a distinct kink in the long fatty acid chains. This means that unsaturated fatty acid chains are unable to pack as closely together as their saturated counterparts. As a result, less energy is needed to disrupt the intermolecular forces that hold unsaturated fatty acids together and they have lower melting points and tend to be liquids at room temperature.

Humans and other mammals are able to synthesize some saturated fatty acids and monounsaturated fatty acids, and can adapt certain fatty acids obtained through their diet by adding two carbon units and introducing double bonds. Fatty acids that can be synthesized in this way are described as non-essential fatty acids. Another set of fatty acids, such as linoleic and linolenic acids (Figure 2.2(b)), are called essential fatty acids. These fatty acids have to be obtained from food stuffs such as vegetable oils, nuts, and seeds: we can't synthesize them ourselves.

Although we cannot synthesize these fatty acids, they are essential precursors for important metabolites, such as eicosanoids. This diverse range of molecules is produced by mammals and has a wide variety of physiological roles. They include prostaglandins that promote inflammation to help fight infection, and the thromboxanes that are produced by platelets to initiate blood clotting. A diet lacking these essential fatty acids can cause poor wound healing, hair loss (alopecia), and dermatitis. There remains much debate about the effects of fatty acids and other lipids on human health, but experimental studies have shown that some are clearly beneficial and some are detrimental to good health. Since we know how much of these compounds are present in different everyday foods, doctors and dieticians can advise about how much—or how little—we should eat of these different types of foods (Table 2.1).

Table 2.1 Sources of fatty acids and lipids in the human diet. Knowledge of the amounts and types of fatty acids in each type of food and their impact on our bodies leads to advice about whether amounts in a typical diet should increase or decrease.

Type of fatty Acid	Examples of food containing them
Saturated fatty acids	e.g. milk and meat products
Trans fatty acids	e.g. hard margarine, vegetable spreads
Omega-3 fatty acids	e.g. oily fish, cod liver oil, soya beans, rape seed oil
Vitamins D & E	e.g. oily fish, fish oils, soya beans/soya oil

nadianb/Shutterstock.com; Ray B Stone/Shutterstock.com; V J Matthew/Shutterstock.com; orientalprincess/Shutterstock.com.

Trigycerides (triacylglycerols)

Fatty acids are able to attach to the three-carbon molecule glycerol by ester bonds (see Figure 2.4). These types of molecules are known as triglycerides (also referred to as triacylglycerols). Since each carbon from the original glycerol molecule attaches to a specific fatty acid molecule, triglycerides can contain a mixture of different fatty acids; these can have different lengths and

Figure 2.4 Ester bonds are key to the formation of triglycerides

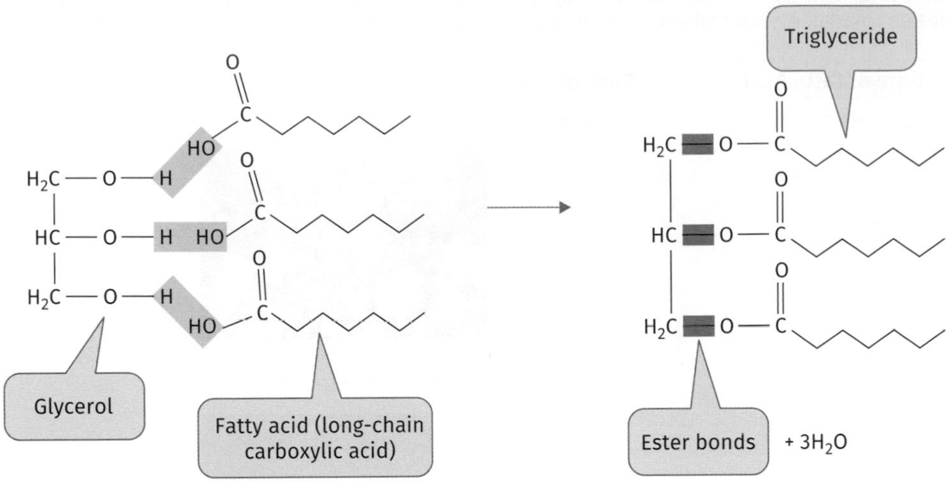

Adapted with permission from Crowe & Bradshaw, *Chemistry for the Biosciences,* third edition. Oxford: Oxford University Press, 2014.

varying levels of saturation. Depending on their fatty acid composition, triglyceride mixtures are referred to as fats or oils: fats are solid at room temperature and oils are liquid. Triglycerides are extremely hydrophobic molecules that come together through their mutual water phobia to create fat droplets. Mammals contain adipose tissue, which is located throughout the body but especially just under the skin and around certain organs—for example, the kidneys. Adipocytes are a specific type of cell, found in this tissue, that are able to store triglyceride droplets. Fat is a poor conductor of heat and adipose tissue is able to slow down and prevent heat loss. Animals can also secrete these fat molecules through specific glands to make feathers and fur water repellent.

Triglycerides are used by organisms to store and transport fatty acids. In fact, they are an incredibly useful way to store energy in cells and tissues because they can be metabolized to release energy. Gram for gram, triglycerides release much more energy than carbohydrates when they are oxidized: 37 kJ is released for each gram of fat that is oxidized compared to 17 kJ for each gram of carbohydrate. In plants, fruits such as avocados and olives, as well as seeds (including sunflower, corn, palm, and coconut) are stuffed with triglycerides, which provide the energy for seed germination and plant growth. Human populations have been exploiting this for centuries by eating the fruits and seeds before they have a chance to germinate! Modern technology means we are now able to extract these plant oils on an industrial scale for our use and consumption.

Isoprenoids

Isoprenoids, such as steroids and terpenes, are biomolecules composed from repeating five carbon units—the isoprene unit (see Table 2.2). Terpenes are found in the essential oils of plants and have been used as perfumes and medicines for centuries. Terpenes formed from two isoprene residues are known as monoterpenes. Three isoprene residues are known as

Table 2.2 The diverse chemical structures of common isoprenoids—along with a common biological example

Common name	Type of terpene	Chemical structure
Isoprene	Hemiterpene (1 isoprene unit)	
Terpene	Monoterpenes (2 isoprene units)	
Farnesene	Sesquiterpenes (3 isoprene units)	
Squalene	Triterpene (6 isoprene units)	
Carotene	Tetraterpene (8 isoprene units)	β-carotene

Chemical structures © Royal Society of Chemistry.

sesquiterpenes. Citronella, an essential oil added to candles as well as soaps and perfumes, contains farnesene, a 15-carbon isoprenoid. Put six isoprene molecules together and you get squalene, also known as a triterpene. This molecule is produced by plants and animals as a precursor to all sterols. These include cholesterol, important for its role in cell membranes among other things, and all of the steroid hormones. The bright orange colour of a carrot comes from the pigment carotene, a tetraterpene composed from eight isoprene units, while natural rubber is formed from 3000–6000 isoprene units. You can see some of these structures in Table 2.2.

Steroids also play important roles in organisms. Steroids have as their basic 'skeleton' the fused hydrocarbon ring shown in Figure 2.5(a). Look at the structure of cholesterol, an example of a steroid, in Figure 2.5(b) and note its fused hydrocarbon skeleton. Cholesterol is an essential molecule in mammalian cells and forms an important part of cell membranes. It is also the starting molecule for a range of secondary messengers, including the sex hormones testosterone and oestrogen. These steroid-derived hormones are able to pass through the hydrophobic cell membrane, enter the cell, and migrate to the cell nucleus attached to a specific hormone receptor protein. The steroid–receptor complex is able to bind to specific nucleotide sequences in DNA and can turn on and up-regulate the transcription of target genes.

Figure 2.5 The structure of steroids. (a) The fused hydrocarbon skeleton of steroids; (b) the structure of cholesterol.

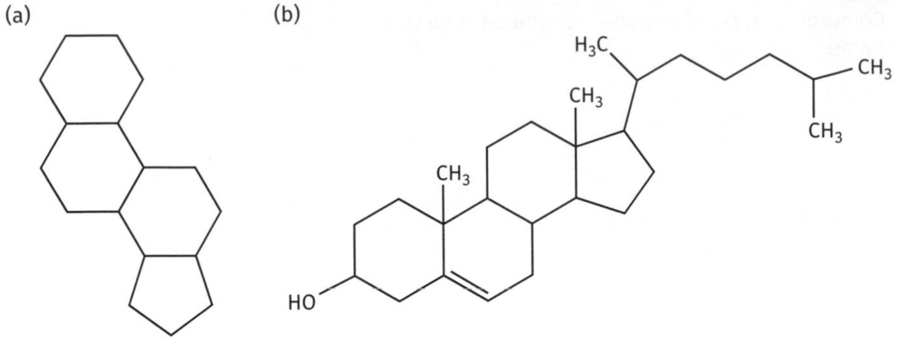

Adapted with permission from Crowe & Bradshaw, *Chemistry for the Biosciences*, third edition. Oxford: Oxford University Press, 2014.

Phospholipids

When it comes to the structure of cells, perhaps the most biologically important class of lipids are the phospholipids. Phospholipids are made from four components: fatty acids, a phosphate group, an alcohol attached to the phosphate, and a 'carbon-based platform' to which the other components are attached (Figure 2.6(a)). An example of a common mammalian phospholipid is phosphatidylcholine (Figure 2.6(b)). The fatty acid components give the phospholipid the ability to form a hydrophobic barrier, whereas the remainder of the molecule has hydrophilic properties, which enables interactions with other types of polar molecules, including water.

Molecules that contain both hydrophobic and hydrophilic regions are said to be amphipathic. They have the important property of being able to adopt particular arrangements when placed in liquids. Phospholipids are perfect examples of amphipathic molecules: when they are placed into water they spontaneously arrange into micelles, or lipid bilayers (Figure 2.6(c)). Micelles are ordered spherical structures where the hydrophobic fatty acid side chains are buried in the middle so that they are excluded from the water and the hydrophilic head groups are orientated to face the water. This property is exploited by cells to form the phospholipid bilayer of cell membranes. Of course, cell membranes are much more than a phospholipid bilayer: many other compounds, including different lipids, carbohydrates, and proteins, enable the membrane to carry out its multiple functions for the cell (Figure 2.6(d)).

Sphingolipids

Phospholipids are not the only lipids found in cell membranes. They also contain sphingolipids, which are composed from long-chain amino alcohols. By comparison with the molecules we have already seen, one of the hydrocarbon chains in a sphingolipid is bound to the lipid via an amide bond, as illustrated in Figure 2.7. In animals, the alcohol component is sphingosine, whereas plants contain phytosphingosine. Sphingomyelin is found in most animal cell membranes and it is particularly abundant in

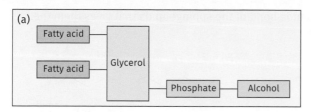

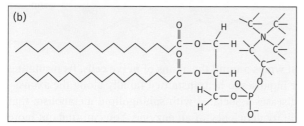

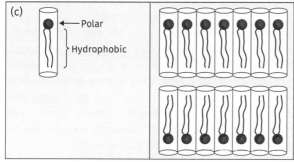

Figure 2.6 Phospholipids and cell membranes. (a) Phospholipids consist of four chemical components: fatty acids, a phosphate group and alcohol attached to it, and a carbon-based 'skeleton' shown as glycerol in this example. (b) Phosphatidylcholine is a common type of phospholipid found in mammalian cells. (c) A simple approximation to the chemical structure of phospholipids is a cylinder with a ball representing the polar end and two wavy lines representing the hydrophobic portion (left-hand image). Since the polar region of the lipids can interact with water but the hydrophobic regions do not like to interact with water, these requirements favour the formation of a lipid bilayer structure (right-hand image). (d) The basic phospholipid bilayer interacts with many other chemicals to produce the familiar structure of the cell surface membrane.

(d)

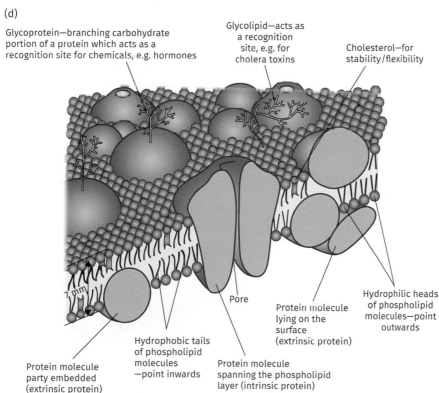

Glycoprotein—branching carbohydrate portion of a protein which acts as a recognition site for chemicals, e.g. hormones

Glycolipid—acts as a recognition site, e.g. for cholera toxins

Cholesterol—for stability/flexibility

Pore

Protein molecule lying on the surface (extrinsic protein)

Hydrophilic heads of phospholipid molecules—point outwards

Protein molecule party embedded (extrinsic protein)

Hydrophobic tails of phospholipid molecules—point inwards

Protein molecule spanning the phospholipid layer (intrinsic protein)

(d): Adapted with permission from Fullick, A. A Level Biology A for OCR Student Book. © Oxford University Press, 2015. Reproduced with permission of the Licensor through PLSclear

Figure 2.7 It is the amide bond of the sphingolipids that gives them their unique properties

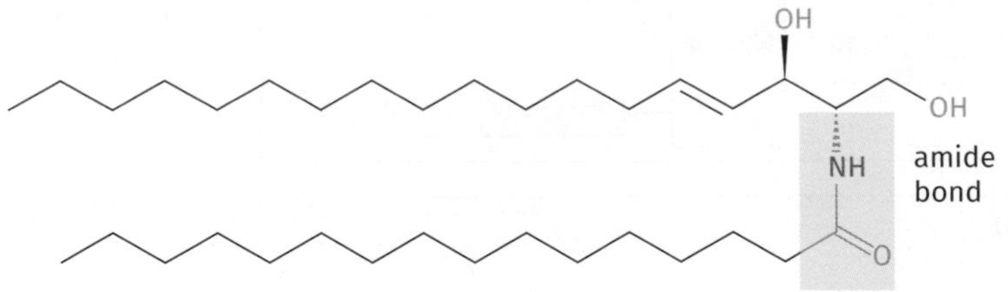

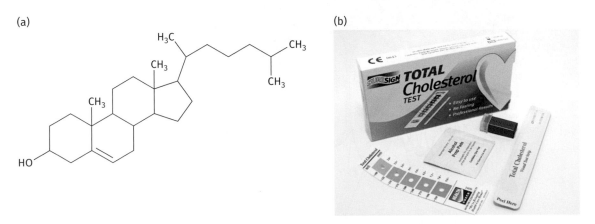

the myelin sheath that wraps around the axons of nerve cells. Its insulating property allows nerve impulses to be transmitted rapidly along the axons.

There are several diseases associated with sphingolipid metabolism that are caused by a hereditary deficiency of an enzyme. You can find out more about one of the most well-known of these, Tay Sachs disease, in some of the further reading highlighted at the end of this chapter.

Membranes also contain one more essential lipid, the steroid cholesterol (see Figure 2.5(b)). Cholesterol is a smaller molecule compared to other molecules in the cell membrane, but it has a vital role to play. For membranes to fulfil their role, they have to provide a barrier that separates the inside of the cell from the environment. At the same time, they have to be permeable and have enough fluidity to allow certain molecules to move through and across this molecular barrier. The molecules within membranes are always moving and this gives membranes the consistency of olive oil.

Just like phospholipids, cholesterol is an *amphipathic molecule*. It has a hydrophilic *and* a hydrophobic region. The hydroxyl (OH) group of the cholesterol steroid aligns with the phosphate heads of the phospholipids and the rest of the molecule becomes embedded into the hydrophobic lipid tail layer (Figure 2.8). Cholesterol decreases the permeability of the cell

Figure 2.8 Cholesterol plays an important part in membrane structures and the production of steroid hormones—but excess levels in the blood can lead to heart disease, so it needs to be monitored

(a): Adapted with permission from Crowe & Bradshaw, *Chemistry for the Biosciences*, third edition. Oxford: Oxford University Press, 2014. (b): Libby Welch/Alamy Stock Photo.

membranes to water. At the same time, it separates the phospholipids by nestling between their hydrophobic tails so that the fatty acid chains can't come together to form rigid crystalline structures. In this way, it ensures the fluidity of membranes is maintained.

Wax esters

Wax esters are complex mixtures of long-chain fatty acids and long-chain alcohols. These waxes form a protective coating on the leaves, fruit, and stem of plants, as well as the skin and fur of animals. In mammals, the principal activity of specialized cells—sebocytes—in mature sebaceous glands is to produce and secrete sebum, a complex mixture of lipids whose composition varies according to the species concerned. Human sebum consists of a mixture of lipids including cholesterol, but the most interesting products are squalene and wax esters. They are unique to sebum and not found anywhere else in the body, nor among the epidermal surface lipids. The sebum of humans is thought to provide protection from light, have antimicrobial activity, deliver fat-soluble anti-oxidants to the skin surface and have both pro- and anti-inflammatory activity. However, the ultimate roles of human sebum, as well as the metabolic pathways regulating its composition and secretion rate, are not fully understood.

Wax esters also form very tough protective coverings for many insects. For example, many scale insects are protected from water loss by a thin film of wax that covers their cuticles (Figure 2.9). Scale insects have modified

Figure 2.9 Waxes are key to the survival of both insects and plants, making them waterproof and protecting them from factors ranging from UV light to pathogen attacks

Anthony Short.

their mouth parts to allow them to pierce and suck up fluids. Many adult females have also lost their legs, and attach to host plants by their modified mouth parts. Unable to move, and locked into position on the host plant, these insects are vulnerable to predators and extremes of temperature. In their defence, scale insects protect themselves with a self-generated waxy coating. This wax also protects their eggs and young.

Proteins: workhorses of the cell

Proteins are essential to all cells: almost all tasks carried out inside cells involve proteins at some point. The variety and complexity of these tasks is astonishing. They range from the transport of molecules, the generation and use of biochemical energy, roles in cellular defence and structural strength, and the catalysis of biochemical reactions. The reason why proteins are so versatile is due to the almost infinite variety of shapes that they can adopt to undertake any task demanded by the cell.

As you will find out in Chapter 3, the information that dictates the structure of proteins is contained within the DNA sequences stored within the cell, known collectively as its genome. The seemingly simple code of A, T, C, G carried by genes is used to determine the sequence of complementary mRNA molecules in a process called transcription. These mRNA molecules are then fed through ribosomes (large complexes of protein and RNA) that 'read' the information stored in the mRNA polymers and translate it into the plethora of different proteins.

What R amino acids?

Just like polysaccharides (see Chapter 1) proteins are polymers, but in this case the monomers are amino acids. There are twenty different, standard, naturally occurring amino acids; each has a unique shape and biochemical properties that are dictated by the atoms they contain. In general, standard amino acids have the same overall structure (Figure 2.10(a)), with a central carbon atom (called the α carbon) attached to:

- a basic amino group, NH_2, which can accept a proton to become positively charged (NH_3^+);
- an acidic carboxyl group, COOH, which can lose a proton to become negatively charged (COO^-);
- a hydrogen atom, H;
- one of twenty different chemical groups, usually referred to as the 'R' group.

When it comes to biology, it is helpful to describe acids and bases using the Bronsted–Lowry theory, which is widely used in chemistry. This theory defines acids as proton (H^+ ion) donors and bases as proton acceptors. Amino acids can act as both of these.

Acid–base reactions always involve acids and bases operating in pairs—what we call a conjugate acid–base pair. One half of the pair is an acid that forms a conjugate base after the reaction, and one is a base that forms a conjugate acid (Figure 2.11). (As an aid to remembering how these are defined, remember that the conjugate base always has one less H atom than the acid, and the conjugate acid has one more H atom than the base.)

Returning now to the amino and carboxyl groups of amino acids, at pH 7 the carboxyl group is in its conjugate base form, $-COO^-$ (it has given up its proton) and the amino group is in its conjugate acid form, $-NH_3^+$ (it has accepted a proton), as shown in Figure 2.10(a). This makes most amino acids neutral at a pH 7. Neutral molecules that have an equal number of positive and negative charges are described as zwitterions; the chemical structures of some common amino acids with this form are shown in Figure 2.10(b).

Since amino acids can react with both acids and bases, they act as natural buffers: they help cells to resist changes in pH.

As is common in nature, there is an exception to the rule, and in the case of the standard chemical structure of amino acids the exception is proline.

Figure 2.10 The chemical structure of standard amino acids. (a) The image on the left shows the chemical composition of an amino acid; the R group varies from amino acid to amino acid, and so gives each amino acid its identity. The image on the right provides an overview of how the different chemical groups are organized in three-dimensional space. In the image on the right, the size of spheres represents the relative size of each chemical group (average size for R group), but does not indicate absolute size. (b) Examples of the varying chemical structures of amino acid R groups, which give them different biochemical properties. Atoms contained within the orange shaded region are the R group of each named amino acid.

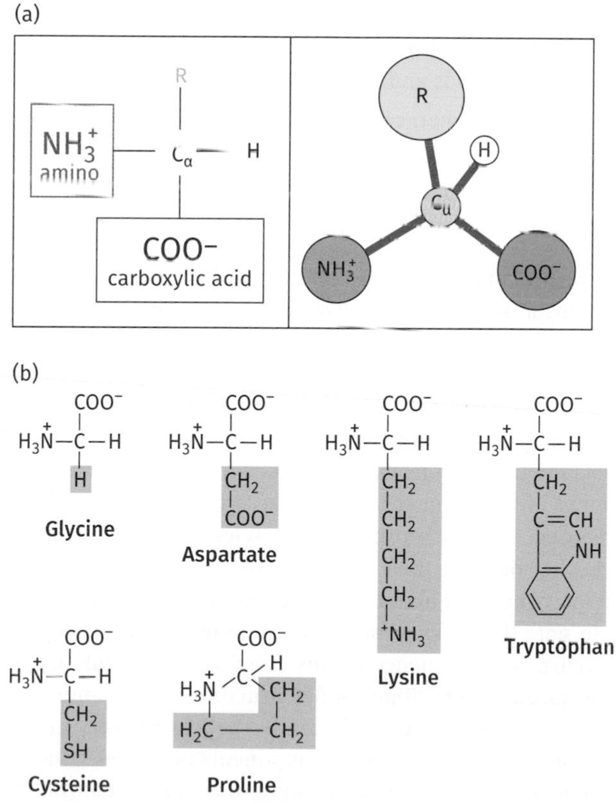

Figure 2.11 The concept of the conjugate acid-base pair

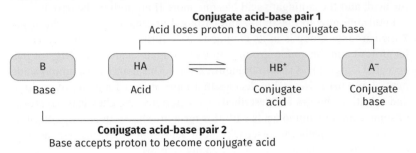

Proline has an imino group formed by a ring closure between its R group and its amino group, giving it a unique shape and structure (Figure 2.10(b)).

The reason there are twenty unique standard amino acids lies in the variety of R groups that they employ. The different R groups can be classified into different types. A common format used by biochemists distinguishes amino acids by their ability to interact with water. As shown in Figure 2.10(b) and Table 2.3, at physiological pH there are amino acids that are:

- **non-polar** and neutral (R groups with no charge and atoms that do not interact well with water and are hydrophobic), such as glycine and proline—some of these R groups also contain aromatic groups, which are relatively large and contain ring structures, such as tryptophan;
- **polar** and neutral (R groups with no charge and atoms that are able to form hydrogen bonds and easily interact with water), such as cysteine;
- acidic (negatively charged R groups that are able to easily interact with water or positively charged chemical groups), such as aspartate;
- basic (positively charged R groups that are able to easily interact with water or negatively charged chemical groups), such as lysine.

The α-carbon of amino acids is a chiral atom because it has four different chemical groups attached (a hydrogen, carboxyl group, amino group, and R group) and has asymmetry (see Figure 2.10(a)). Interestingly, glycine has a second hydrogen as its R group and is not asymmetrical—another biological exception to a general rule. Consequently, the α-carbon in glycine is not chiral.

As already highlighted for saccharides in Chapter 1, molecules with chiral or asymmetric carbons can exist as **stereoisomers**: the arrangement of their atoms in three-dimensional space are mirror images and cannot be superimposed over each other. When scientists talk about isomers of amino acids, they describe them as D- or L-amino acids. Although we do not need to be too concerned about the differences between these isomers, it's important to note that only L-amino acids are used during the synthesis of natural proteins. A consequence of this chirality is that polypeptide chains usually form specific three-dimensional structures—for example, they often twist into right-handed helices. Left-handed helices are very rare in nature.

Although amino acids are the key components of proteins, they also have other essential roles in cells. For example, tryptophan can be chemically

Table 2.3 Biochemical characteristics of amino acids.

Amino acid	3-letter abbreviation	Biochemical characteristics
Alanine	Ala	Neutral, hydrophobic
Arginine	Arg	Basic, positively charged
Asparagine	Asn	Neutral, polar
Aspartic acid	Asp	Acidic, negatively charged
Cysteine	Cys	Neutral, polar
Glutamine	Gln	Neutral, polar
Glutamic acid	Glu	Acidic, negatively charged
Glycine	Gly	Neutral, hydrophobic
Histidine	His	Basic, positively charged
Isoleucine	Ile	Neutral, hydrophobic
Leucine	Leu	Neutral, hydrophobic
Lysine	Lys	Basic, positively charged
Methionine	Met	Neutral, hydrophobic
Phenylalanine	Phe	Neutral, hydrophobic
Proline	Pro	Neutral, hydrophobic
Serine	Ser	Neutral, polar
Threonine	Thr	Neutral, polar
Tryptophan	Trp	Neutral, hydrophobic
Tyrosine	Tyr	Neutral, polar
Valine	Val	Neutral, hydrophobic

modified to form serotonin and melatonin, neurotransmitters released by nerve cells. Tryptophan can also be chemically modified to form thyroxine, a hormone produced by the thyroid gland of animals. Amino acids are also the precursor molecules of the nitrogenous base components of nucleotides and nucleic acids, as well as of iron-containing haem groups and magnesium-containing chlorophyll.

Sometimes, standard amino acids can be modified after a polypeptide has been synthesized. Collagen is one of the most common protein molecules found in connective tissue and it contains significant amounts of 4-hydroxyproline and 5-hydroxyproline (Figure 2.12). These modified proline molecules are important structural components of this key fibrous protein.

Figure 2.12 The structures of 4-hydroxyproline and 5-hydroxyproline as compared to 'standard' proline. Notice how the carbon atom to which the hydroxyl (–OH) group is attached differs between 4- and 5-hydroxyproline.

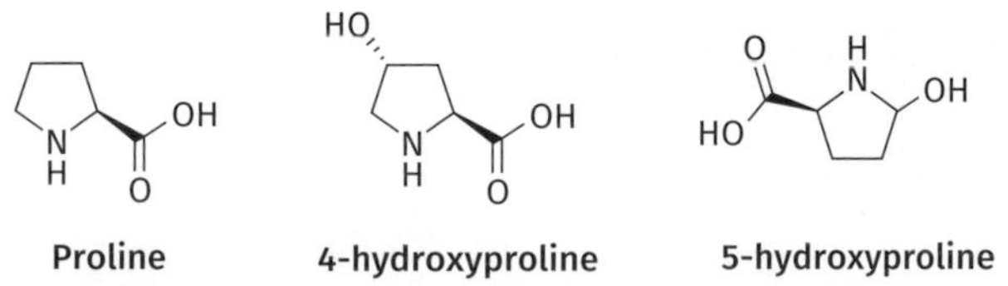

Proline   4-hydroxyproline   5-hydroxyproline

Polypeptides are formed when bonds—called peptide bonds—form between the amino group of one amino acid and the carboxyl group of another. The formation of these bonds results from a condensation reaction that releases H_2O (Figure 2.13). When amino acids have undergone condensation, with a peptide bond forming between them, they become properly referred to as amino acid *residues*.

Polypeptides are usually written with the amino acid residue that has the free amino group on the left; this is called the N-terminal amino acid. The amino acid residue with the free carboxyl group is written on the right; this is the C-terminal amino acid. The peptide bonds that form between amino acids are flat (planar) and rigid, which leads to specific three-dimensional shapes being favoured. This structure and biochemistry of the specific R groups generates the biologically active polypeptides.

Figure 2.13 Formation of the peptide bond. Polypeptide chains are made when a covalent bond—the peptide bond—is formed between the amino group of one amino acid and the carboxyl group of another, releasing H_2O in the process. The size of spheres represents the relative size of each atom or chemical group. The blue shading denotes the flat, rigid peptide bond.

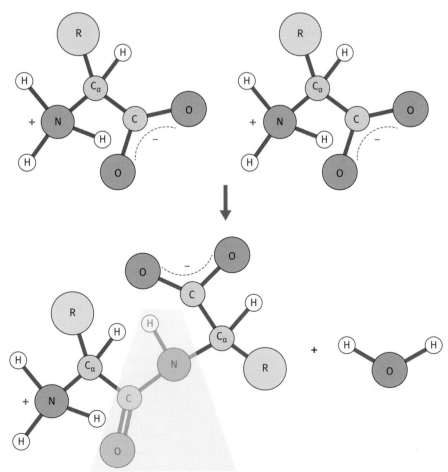

Another set of biochemical reactions has an important impact on the structure of proteins. This leads to bonds being formed between the R groups of two cysteines. Cysteine is an amino acid with a particular reactive biochemical group, the thiol or sulfhydryl group, –S–H (Figure 2.10(b)). When this group is oxidized a disulfide bond or a disulfide bridge (R_1–S–S–R_2) can form between two neighbouring cysteines, as depicted in Figure 2.14. These covalent biochemical bridges can form between cysteines in a single polypeptide chain and between cysteines found in two separate polypeptides. They play an important role in locking down and stabilizing the biologically important three-dimensional shape of proteins.

Though there are many different proteins, they fall into two main classes. Fibrous proteins such as collagen are long, rod-shaped molecules that are insoluble in water. In contrast, globular proteins are compact and often water-soluble. If proteins contain amino acids only, they are known as simple proteins. But if they contain additional biochemical groups such as carbohydrates (glycoproteins), lipids (lipoproteins), haem groups (haemoproteins), or metal ions (metalloproteins), they are described as conjugated proteins.

Protein shape and structure

The ability of each protein to perform its specific function is determined by the unique sequence of amino acids that make up the polypeptide chain (or chains) from which it is formed. These amino acids interact with each other, twisting and bending the single-stranded polymer into unique, complex, three-dimensional shapes.

Primary structure

Proteins are incredibly complex molecules with diverse shapes and functions. This diversity is a consequence of several layers of structural organization. First of all, there is the primary structure, the specific

Figure 2.14 The formation of a disulfide bridge between the thiol groups of neighbouring cysteines in a polypeptide

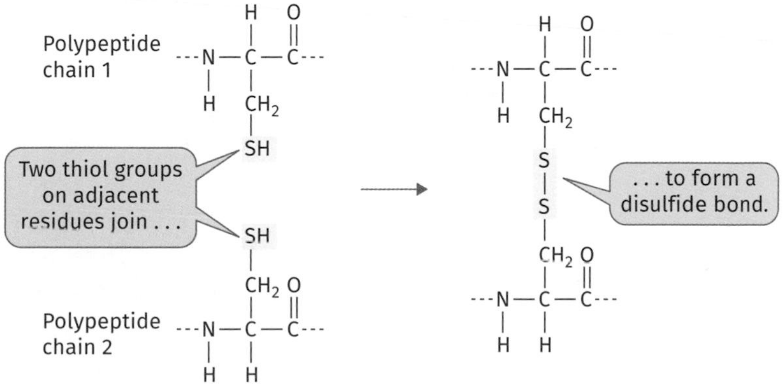

Adapted with permission from Crowe & Bradshaw, *Chemistry for the Biosciences*, third edition. Oxford: Oxford University Press, 2014.

sequence of amino acids as you move along the polypeptide chain from N– to C-terminus. The primary structure–the identity of the amino acids at each point along the polypeptide–is dictated by the genetic information encoded by the gene for that polypeptide.

Secondary structure

Moving beyond their primary structure, proteins take on secondary structures that occur as a result of the polypeptide backbone taking on certain characteristic structural conformations, patterns, and shapes. These characteristic shapes result from the way primary structures feature certain repeating patterns of amino acids. These patterns in turn mean the polypeptide forms certain shapes and structures over others.

Two of the most common secondary structures are the α-helix and the β-pleated sheet; these structures are illustrated in Figure 2.15. The α-helix is a rigid, rod-shaped helix that is formed when the N–H group of

Figure 2.15 The secondary structure of proteins often involves alpha-helices or beta-pleated sheets. These structures form as a result of hydrogen bonds between the amino and carboxyl groups of amino acids in a polypeptide chain. (a) An alpha-helix. (i) and (ii) show cartoon depictions of the three-dimensional structure, while (iii) shows how the alpha helix is stabilized by a network of hydrogen bonds forming between amino (–NH) and carboxyl (–C=O) groups along the polypeptide chain. (b) A beta-pleated sheet. (i) and (ii) show carton depictions of this structure, while (iii) shows the network of hydrogen bonds that stabilizes it. (c) A Greek key motif, evoked by the shape of an antiparallel beta sheet.

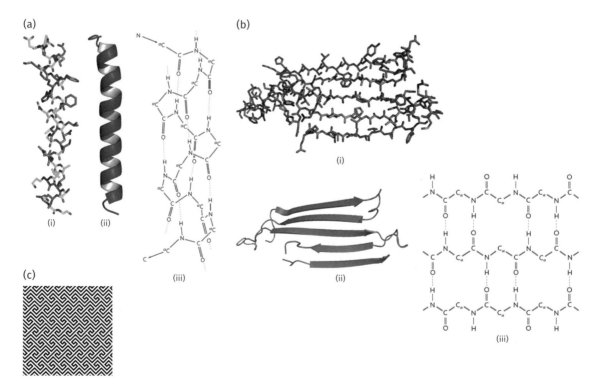

(a) and (b): Adapted with permission from Crowe & Bradshaw, *Chemistry for the Biosciences*, third edition. Oxford: Oxford University Press, 2014. (c): ALMAGAMI/Shutterstock.com.

one amino acid forms a hydrogen bond with an oxygen of a carboxyl group of an amino acid that is four residues away. The R groups of the amino acids point outwards from the helix like petals on a flower. Certain R groups are much more inclined to be involved in this type of structure than others. For example, alanine and leucine are often found in an α-helix, whereas other R groups that are charged (lysine and aspartate, for example) or are just too bulky (tryptophan) are not found in this type of conformational pattern.

β-pleated sheets form when two or more sections of the polypeptide chain line up side by side and hydrogen bonds form between the polypeptide backbone and the oxygen of carboxyl groups of adjacent chains. β-pleated sheets can be parallel, where the polypeptide chains run in the same direction, from N- to C-terminus, or they can be antiparallel—the polypeptide strands run in opposite directions. Antiparallel β-pleated sheets occur when the polypeptide chain doubles back on itself to resemble a pattern seen on ancient pottery, a Greek Key.

Fibrous proteins such as α-keratin, collagen, and silk contain a high proportion of regular secondary structures. Keratin contains α-helices and silk consists of β-pleated sheets (see Figure 2.15), while collagen contains three polypeptide strands that are intertwined within a left-handed helix.

Tertiary structure

The third level of structural organization exhibited by proteins is called the tertiary structure. The tertiary structure arises from the folding-up of secondary structures into more compact shapes. Some examples are shown in Figure 2.16. The tertiary structure is stabilized by interactions between amino acid side chains (the R groups), which hold the structure in its most energetically favourable state. For example, a particular tertiary structure may bring hydrophobic side chains into close proximity, forming a stable structure from which water is excluded; a tertiary structure may also be stabilized by electrostatic interactions occurring between ionic groups of opposite charge (e.g. lysine and aspartate, as shown in Figure 2.10(b)). Further, it may be stabilized by hydrogen bonds that form between polar amino acids side chains or between polar amino acid side chains and the peptide backbone. Finally, covalent bonds can also form along the polypeptide chain, with perhaps the most well-known of these being the disulfide bridges described in the section 'What R amino acids?'.

Quaternary structure

Many proteins, including well-known, essential molecules such as haemoglobin and antibodies, are composed of more than one polypeptide chain. Such sets of polypeptides are held together by a suite of different biochemical interactions including hydrogen bonds, hydrophobic interactions, and covalent cross-linking. When this type of organization occurs for a protein it is referred to as its quaternary structure (Figure 2.17). In addition, the individual interactions between the different subunits are often affected by the binding of ligands, as seen in proteins such as myoglobin and haemoglobin that bind to an iron-containing haem group or ligand.

Figure 2.16 Examples of tertiary structures of proteins. (a) This tertiary structure is composed solely of alpha-helices. (b) The membrane protein FhuA. Notice how the beta strands form a barrel structure. (c) The enzyme triosephosphate isomerase, with a beta-barrel at the centre of the molecule, and alpha-helices around it. (d) The enzyme dihydrofolate reductase, with a beta-sheet running down its interior.

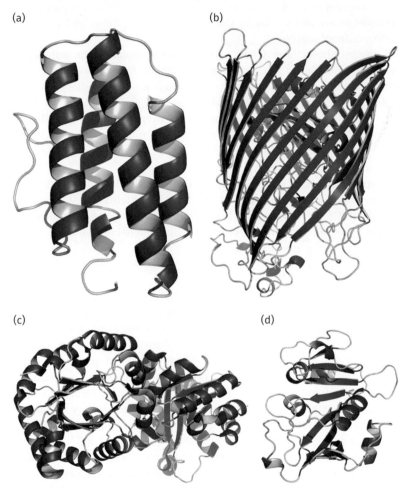

(a)

(b)

(c)

(d)

Adapted with permission from Crowe & Bradshaw. *Chemistry for the Biosciences*, third edition. Oxford: Oxford University Press, 2014.

Proteins as enzymes

Arguably one of the most important roles of proteins is their ability to act as catalysts: they are biological machines that enable biochemical reactions to happen with enough precision and speed to maintain life. We can represent some chemical reactions in the form X+Y = Z; such reactions are driven by energy that allows chemical bonds to be broken and new bonds to form. In a test tube that energy is often supplied as heat, but in living cells the addition of heat can have catastrophic effects: it can damage biological molecules, including proteins, beyond repair. Living systems have developed biological solutions to enable biochemical reactions at the rate

Figure 2.17 Quaternary structure of proteins. This image shows the structure of ATP synthase, which comprises numerous individual polypeptides, each of which is depicted in a different colour here.

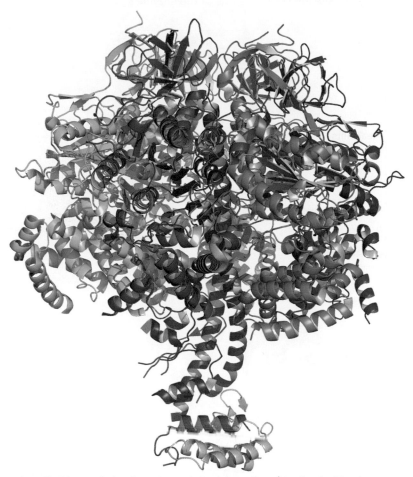

Adapted with permission from Crowe & Bradshaw, *Chemistry for the Biosciences*, third edition. Oxford: Oxford University Press, 2014.

required for life to be maintained to take place: enzymes! In Chapters 4 and 5, we will explore the ways that some enzymes participate in metabolism to take advantage of the energy within cells.

Enzymes are amazing machines. They are highly specific to the reactions they catalyse and enable them to happen at a rapid rate without altering the intrinsic biochemistry of the reaction. How do they do this? Every biological reaction has what is known as an activation energy—an amount of energy that must be made available to the system for the reaction to proceed (e.g. to allow a reactant (or reactants) to be converted to a product, or products). Generally speaking, enzymes *decrease* the activation energy that is required for biological reactions to take place. If less energy is needed to make them go, the reactions happen more readily—which we observe as them happening more rapidly. Figure 2.18 shows you the impact of an enzyme on the activation energy required for a reaction to happen.

Figure 2.18 Enzymes as catalysts. (a) Enzymes work by reducing the activation energy required for a reaction to take place, making it possible within a biological system. (b) The specific shape of the protein molecule and its active site are key to the role of the enzyme.

(a)

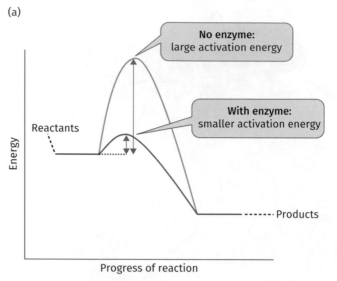

(b)

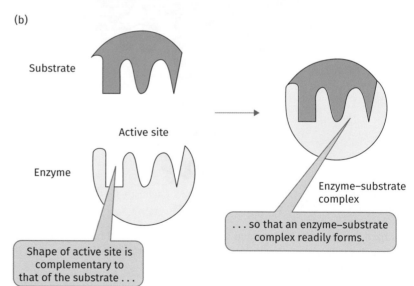

Adapted with permission from Crowe & Bradshaw, *Chemistry for the Biosciences*, third edition. Oxford: Oxford University Press, 2014.

Each individual enzyme has an active site—a small pocket or patch of amino acids and their individual side chains—that contains a binding site that is able to fit around the reaction substrates (Figure 2.17(b)). As the substrates bind to the enzyme, their chemical structures are altered, perhaps bringing them into closer contact or pushing them towards the structures of the products. Once lined up correctly, a slight change in the conformation of the enzyme can result in the formation of a strained enzyme–substrate complex that makes it easier for the reaction to take place.

Although some enzymes rely solely on amino acid side groups to bind the substrate and catalyse reactions, others are conjugated proteins: they rely on additional non-protein chemical components such as Mg^{2+} and Zn^{2+} (called 'cofactors') or complex organic molecules such as haem groups (called 'coenzymes') for their biochemical activities. Amino acid side chains and cofactors, such as metal ions, can also catalyse proton transfer and may be useful as electrophiles, allowing redox reactions to take place.

These biochemical properties allow enzymes to catalyse an array of biochemical reactions—but these reactions can actually be grouped into six major categories according to the type of enzyme involved; these six categories are shown in Table 2.4.

The activity of enzymes can be inhibited by molecules that reduce their ability to undertake their roles as biological catalysts; such molecules can slow down or even completely turn off the enzymes they inhibit. Inhibitors have an important role in living systems. They make it possible to regulate the rate of certain biochemical reactions that are part of metabolic pathways, which in turn means cells, tissues, and whole organisms can respond to different demands.

Table 2.4 Classes of enzymes and the reactions they catalyse

Enzyme class	Reaction catalysed	Typical reaction	Example names of enzyme
Oxidoreductase	Oxidation–reduction reactions	$AH + B \rightarrow A + BH$ (reduced) $A + O \rightarrow AO$ (oxidized)	Oxidase Oxygenase Reductase Peroxidase Hydroxylase
Transferases	Transfer of chemical group from one molecule to another	$AB + C \rightarrow A + BC$	Transaminase Kinase
Hydrolases	Cleavage of chemical bonds by adding water	$AB + H_2O \rightarrow AOH + BH$	Esterase Phosphatase Peptidase
Lyases	Removal or addition of chemical groups without use of water; usually removed from or added to a double bond	$RCOCOOH \rightarrow RCOH + CO_2$ $[X–A + B–Y] \rightarrow [A=B + X–Y]$	Decarboxylase Deaminase Lyase
Isomerase	Intramolecular rearrangements, including the inversion of asymmetric carbon atoms, and the intramolecular transfer of functional groups	$ABC \rightarrow BCA$	Isomerase Mutase
Ligases	Formation of bonds between two substrate molecules; relies on energy supplied by hydrolysis of a co-substrate, such as ATP	$X + Y + ATP \rightarrow XY + ADP + P_i$	Carboxylase Synthetase

In addition, modern medicine relies on many clinical drugs that act as enzyme inhibitors. Antibiotics are a classic example. Drugs such as penicillin covalently bind to a specific enzyme that a pathogenic bacterium relies on to build and maintain its cell wall. This enzyme—a transpeptidase—helps to construct the peptidoglycan cell wall of the bacterium. When penicillin binds to it, however, the enzyme is no longer able to construct and repair a viable cell wall. As a result, the bacterium cannot maintain its biochemical integrity, the weak cell wall ruptures, and the bacterium dies.

Any environmental factor that interferes with or disturbs the intrinsic structure of the protein molecule will also interfere with the ability of the enzyme to catalyse its target biochemical reaction. This includes changes to pH and temperature. Every enzyme has an optimal temperature, which is usually close to that of its usual environment. For example, human enzymes usually function best at body temperature, 37°C. If the temperature is raised significantly above this level, the protein denatures and enzymatic activities are negatively affected. (However, some enzymes can operate at quite extreme temperatures, as we discover in Case study 2.1.) Similarly, enzymes only perform well in a narrow pH range. Compromising the environmental pH range can also have a severe impact on the integrity and catalytic activity of the enzyme.

Case study 2.1
'Hot enzymes'

Although many animals, including humans, rely on a temperature of 37°C to sustain life, a range of other life forms have adapted to live at much higher temperatures: the thermophiles. Thermophiles thrive in temperatures between 41 and 122°C; they include archaea and bacteria. They can be found near geothermal sites on land and in oceans, as well as on decaying plant matter found in compost heaps and peat bogs.

Soon after their discovery, scientists began to realize that these organisms could be used in interesting and important ways. They reasoned that if microbes were able to thrive in these thermal environments, then perhaps the enzymes they contained would be stable at higher temperatures too. Scientists began to search for microbes from thermal environments in order to find enzymes that could significantly increase the temperature range for enzymatic bioprocesses. One of the early successful commercialized examples was the discovery, isolation, and production of a thermostable enzyme, *Taq*-polymerase. This enzyme was purified from the thermophilic bacterium *Thermus aquaticus* and is used in **polymerase chain reactions (PCR)** through which DNA is 'amplified' (whereby many copies of a DNA sequence of interest are rapidly synthesized). This process uses high temperatures to melt the double-stranded DNA and to keep the individual strands apart while a DNA primer is added and hybridized to the individual DNA strands.

The *Taq*-polymerase then uses the primer as the starting point to synthesize new strands of double-stranded DNA. You can find out more about PCR and the way it is used in Chapter 3.

A number of other enzymes from thermophilic sources have also been commercialized. These include enzymes that catalyse the breakdown of cellulose, hemicellulose, and lignin found in plants. These thermostable enzymes offer robust alternatives to the use of potentially hazardous chemicals. They are able to withstand the often relatively harsh conditions of industrial processing used to break down cellulose, hemicellulose, and lignin, producing smaller molecules that can be used to make products such as cellophane and viscose.

Proteins as receptors

Protein molecules have many roles in cells in addition to acting as catalysts. For example, they act as membrane receptors, which are vital for the metabolism of all living organisms. Receptors are part of the mechanism cells use to monitor and respond to changes in the environment (see Case study 2.2), and allow cells to recognize and interact with each other. Receptor proteins sit in the lipid membrane of cells, connecting the inside of the cell to the outside environment. As such, they are classified as integral membrane proteins.

In general, receptor proteins have many α-helices and β-pleated sheets, and the polar side chains are tucked inside the protein molecule, leaving the amino acids with hydrophobic, lipid-loving side chains on the outside of the protein molecule where they make contact with the lipid bilayer of the cell.

One such protein receptor is rhodopsin, which is found in the rod receptors at the back of our eyes. Rhodopsin is a reddish-purple protein that absorbs green-blue light. It is sometimes referred to as visual purple and is responsible for *monochromatic* vision in the dark. Rhodopsin is made up of two components, a protein known as scotopsin and a covalently bound cofactor known as retinal, which is produced in the retina from vitamin A and beta carotene. Scotopsin is an opsin, a light-sensitive membrane receptor that crosses the cell membrane using seven protein transmembrane domains. These domains form a pocket where retinal lies horizontally to the cell membrane. This retinal is linked to a lysine amino acid residue found in the seventh transmembrane domain of the opsin protein. This structure is illustrated in Figure 2.19(a). Thousands of rhodopsin molecules are found in each outer segment disc of the host rod cell.

The absorption of green-blue light by the rhodopsin sets off a series of conformational changes ('bleaching'). These conformational changes activate a series of chemical messengers inside the cell that eventually allow our brains to recognize and detect light (see Figure 2.19(b)). Central to these conformational changes is the way retinal switches between being a *cis*- and being a *trans*-isomer, as discussed in Chapter 1.

Figure 2.19 The roles of protein receptors in human vision. Sophisticated biochemical changes in rhodopsin, whose structure is shown in (a), make our vision possible via the mechanism shown in (b)

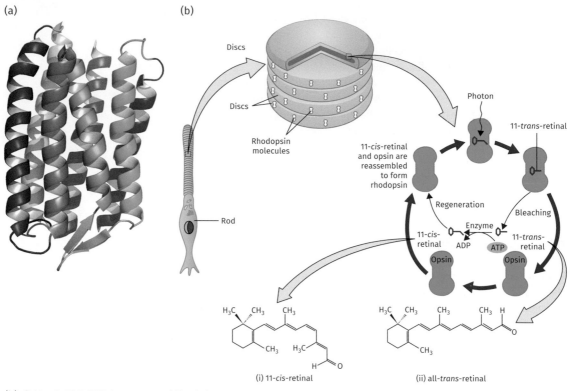

(a)

(b)

Discs

Discs

Rhodopsin molecules

Rod

Photon

11-*trans*-retinal

11-*cis*-retinal and opsin are reassembled to form rhodopsin

Regeneration

Bleaching

Enzyme

11-*cis*-retinal   ADP   ATP   11-*trans*-retinal

Opsin   Opsin

H₃C CH₃ CH₃

CH₃ H₃C

H O

(i) 11-*cis*-retinal

H₃C CH₃ CH₃ CH₃ H

CH₃ O

(ii) all-*trans*-retinal

(b): © Mar 3, 2014 CNX Anatomy and Physiology.

Case study 2.2
Migration, magnets, and cryptochrome receptors

One of the many mysteries that has puzzled scientists is how birds, such as the robin (Figure A), and some other animals can use a so-called 'sixth sense' to navigate entire continents on their long annual migrations. Scientists believed that this navigation—a behaviour where the animal reliably prefers to travel in a certain direction—takes advantage of geomagnetic lines that run from the north to south pole of the Earth. Interestingly, birds do not distinguish between 'magnetic North' and 'magnetic South', which are the readings of a compass based on polarity. Instead, they distinguish between 'poleward', where the field lines are pointing from the equator towards a pole, and 'equatorward', where they point from a pole towards the equator. This ability is referred to by scientists as 'magnetoreception'.

How birds such as the European robin and zebra finches can sense these magnetic fields has remained a mystery for many years. However, scientists have

Figure A Robins are such familiar birds—yet we don't yet fully understand how they use magnetic fields to navigate

Anthony Short.

started to uncover clues that suggest that the sensors used by birds to sense these magnetic lines are protein receptor molecules called cryptochromes (CRY1, CRY2, and CRY4), which are found in cones at the back of their eyes. One of these proteins acts as a light-dependent, radical-pair-based magnetic compass. But what exactly does that mean? The explanation starts with the requirement for the magnetic compasses of birds to absorb photons of light in the short-wavelength part of the light spectrum (373 nm UV, 424 nm blue, 502 nm turquoise, and 565 nm green). Cryptochromes are blue-light photoreceptors that can absorb the energy available in UV and blue light due to the presence of flavin molecules that are embedded within the globular protein sheaths. These flavin molecules are essential for the cryptochromes to act as light-dependent magnetoreceptors (Figure B).

Cryptochromes were initially identified as being involved in the regulation of **circadian rhythms**. Scientists studying cryptochromes in zebra finches found that CRY1 and CRY2 did what would be expected of substances that regulate circadian rhythms—their levels fluctuated depending on the time of day. But when they examined CRY4, its levels remained constant, suggesting this protein had another role that required it to be active regardless of the time of day.

A different group of scientists compared the levels of cryptochromes produced by European robins during their migratory and non-migratory seasons, and by domestic chickens that do not migrate. Unlike CRY1 and CRY2, the level of CRY4 in European robin retinae is significantly higher during the migratory season compared to the non-migratory seasons, but a similar increase isn't seen in domestic chickens which do not undertake an annual migration. This evidence suggests that the cryptochrome most likely to be involved in magnetoreception is the CRY4 protein. However, more experiments are needed to fully understand this theory.

Figure B Migratory birds such as this wheatear (a) depend on the structure of cryptochromes (b) to give them the magnetic sense they rely on for magnetic navigation from Africa to the UK and back every year. Cryptochromes use flavin molecules (FAD) that are embedded within the protein (shown here in orange) to absorb the energy available in UV and blue light.

(a)

(b)

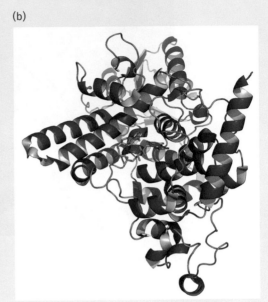

(a): Anthony Short.

Chapter summary

- Lipids are a diverse group of naturally occurring substances that perform many important roles in living organisms. Different chemical groups are attached to different lipids, leading to varying levels of solubility in water, but they are all able to dissolve in non-polar solvents.
- Lipids are essential components of cell membranes, they can be used to store energy, they are able to provide a waterproof coating to animals, and they are the basic building blocks of essential molecules that include hormones, vitamins, and pigments.
- Lipids are classified in different ways and they include fatty acids, phospholipids, and triglycerides.
- Proteins are essential to all cells since almost all tasks they perform involve proteins at some point. The variety and complexity of these tasks ranges from transporting molecules, or generating and using

biochemical energy, to providing cellular defence and structural strength, as well as catalysing biochemical reactions.

- Proteins are polymers made from twenty different, naturally occurring, standard amino acids. These have a similar overall structure, with a central carbon atom attached to an amino group, a carboxyl group, a hydrogen atom, and one of twenty different chemical groups, usually referred to as the 'R' group or side chain.

- The versatility of proteins is due to the great variety of shapes that they can adopt. Overall structures are dictated by their primary structure, the sequence of amino acids, but these take on different secondary structural conformations, patterns, and shapes. These come together to form three-dimensional conformations, or tertiary structures, that are unique to each protein. Some proteins are composed of more than one polypeptide chain and this type of organization is referred to as quaternary structure.

Further reading

https://ghr.nlm.nih.gov/condition/tay-sachs-disease.

https://www.genome.gov/10001220/learning-about-taysachs-disease.

https://pdb101.rcsb.org/motm/147: a useful online resource providing background information about rhodopsin.

Discussion questions

2.1   Many diets suggest cutting back on fat or even eliminating fat from our diet. Discuss reasons for this advice—and why it might not be such a good idea.

2.2   What structural phenomena might 'hot proteins' use to ensure their stability at high temperatures?

3 NUCLEOTIDES AND NUCLEIC ACIDS: BIOLOGY'S INFORMATION STORES

Nucleotides and **nucleic acids** are some of the best-known biological molecules. They exist in all cells and have been widely studied. Nucleotides play central roles in supplying the energy for many metabolic reactions, and in the storage and use of genetic information: they are key players in passing on genetic material from one generation of cells to the next. They are also involved in controlling how the genetic information is used—e.g. they play an important part in switching genes on or off and so determining which proteins are synthesized.

Although there are many different nucleotides, they all consist of a nitrogen-containing base, a five-carbon sugar, and one or more phosphate groups. **Adenosine-5′-triphosphate**, or **ATP**, is the nucleotide you are most likely to have heard of.

Nucleic acids are polymers of nucleotides. There are two types, **ribonucleic acid (RNA)** and **deoxyribonucleic acid (DNA)**. DNA molecules usually form elegant, double-stranded molecules, that make the famous 'double helix', and they are the genetic material in most cells (Figure 3.1). RNA molecules are single-stranded, and include messenger RNA (mRNA), transfer RNA (tRNA), ribosomal RNA (rRNA), and small, regulatory RNAs. All of these RNA molecules are involved in protein synthesis—the process by which the genetic information in DNA is used to guide the synthesis of polypeptides, which go on to have a role in specific cellular structures or activities.

As scientists' understanding of DNA and RNA molecules developed, the 'era of **molecular biology**' and, more recently, **synthetic biology**, emerged. Genetic engineering and gene editing techniques allow the manipulation of genes into novel forms and are fundamental to **biotechnology**, where biological solutions are applied to some of the major challenges facing humanity.

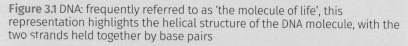

Figure 3.1 DNA: frequently referred to as 'the molecule of life', this representation highlights the helical structure of the DNA molecule, with the two strands held together by base pairs

encg/Shutterstock.com.

Basic nomenclature and chemical structures

Nucleotides have incredibly important biochemical roles in most processes that occur in living organisms. Although they are well known as the basic building blocks of the genetic code, they also provide energy that can be used to fuel enzyme-catalysed reactions. The nucleotide that is best known for this is adenosine-5'-triphosphate or ATP (see Figure 3.2), which acts a universal currency of energy in biological systems. ATP provides energy to cells because the chemical products released when the phosphate groups are hydrolysed are more stable than ATP. This means that the hydrolysis reaction is highly favoured in cells. In fact, ATP is so important that humans synthesize approximately their own mass in ATP every day!

As you can see in Figure 3.2, nucleotides have three biochemical components, with the following chemical structures:

1. **Base:** there are two types of nitrogen-containing ring compounds, purines and pyrimidines. Purines contain two carbon–nitrogen rings, as shown in Figure 3.2, while pyrimidines have only one (see Figure 3.3). This size difference affects the space taken up by the base, which is crucial to the structure of DNA (see Case study 3.1). Various chemical groups are joined at different positions to the ring atoms, and this is how the different types of base are formed.

Figure 3.2 Chemical structure of adenosine-5′-triphosphate (ATP), a nucleotide. All nucleotides consist of a base, a sugar, and a phosphate ester. These constituent parts are shown for ATP, where the base is adenine (shown in green), the sugar is ribose (shown in purple), and the phosphate is triphosphate (shown in orange). The numbering of the carbon atoms within the sugar follows a regular pattern. Bond lengths are not drawn to scale.

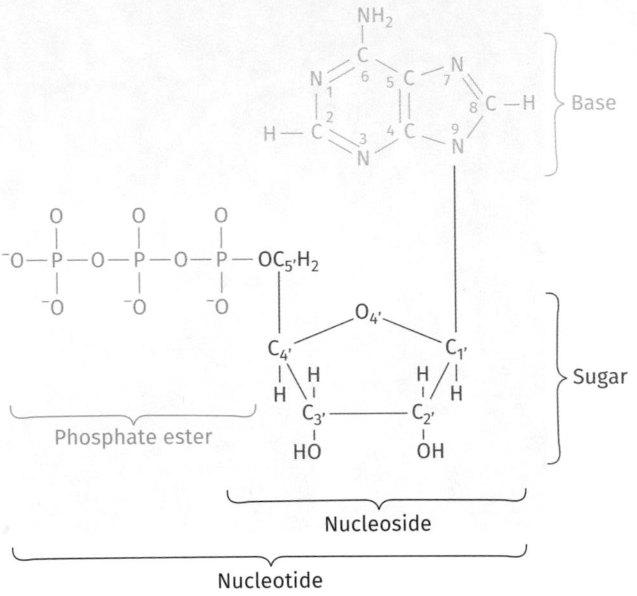

2. **Sugar:** a five-carbon sugar is linked covalently to the base. This sugar is usually ribose, as shown in Figure 3.2, or deoxyribose (see Figure 3.3), which have five carbon atoms, numbered 1′ to 5′. The sugars are locked into a ring structure by a covalent bond that links the C1′ atom of the sugar to the nitrogen-containing base. You may have wondered why scientists talk about 'ribo–' or 'deoxyribo–' molecules. This is because ribose contains a hydroxyl group at C2′, but deoxyribose only has a hydrogen attached to the C2′—it is without oxygen, or deoxy!

3. **Phosphate ester:** phosphate groups are attached to the sugar through covalent chemical bonds known as ester bonds. The phosphate groups usually attach to the ribose ring via the hydroxyl group at the carbon C5′. One, two, or three phosphates are joined, producing mono-, di-, and triphosphates, respectively.

Nucleosides have very similar structures to nucleotides, but only contain a base bound to a sugar—the phosphate groups are missing. In the majority of situations, the chemical structures of nucleosides and their related nucleotides are similar. You have to be careful both writing and reading these names, to make sure you are referring to the right type of molecule!

Scientists use a complex terminology to describe nucleotides and their related compounds. We use standard definitions to help understanding; an

Figure 3.3 Chemical structure of DNA and base pairs. The nomenclature for each base is indicated. Atoms are numbered for one sugar, one purine base, and one pyrimidine base. A single phosphodiester linkage is shown between adjacent nucleosides on each strand. Arrows highlight the orientation of each polynucleotide strand in a duplex—one runs 5' to 3', and the other 3' to 5'. This arrangement is called an antiparallel orientation. Note that bond lengths are not drawn to scale and some have been exaggerated for clarity. Broken lines represent hydrogen bonds in each base pair.

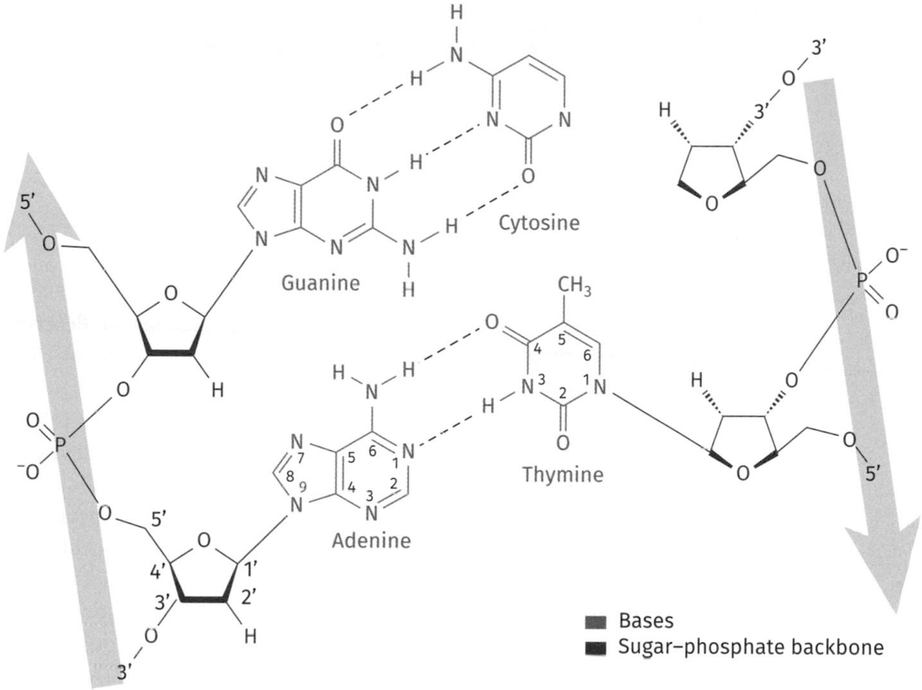

overview of general terms that are used is given in Table 3.1. Hopefully you will find these helpful!

As you can see in Table 3.1, nucleotides are usually referred to in abbreviated terms. If we focus on one of the nucleotides for example, molecules with respectively one, two and three phosphates attached to adenosine are AMP (adenosine **mono**phosphate), ADP (adenosine **di**phosphate), and ATP (adenosine **tri**phosphate); whereas the deoxy variants that are incorporated into DNA are dAMP, dADP, and dATP. Other nucleotides are referred to in similar ways although, confusingly, there is an exception to the rule: thymidine and deoxythymidine are often used interchangeably, and the prefix ribo- is used for ribonucleotides of thymine, instead.

Types of nucleic acids and their structures

Nucleotides form polymers, called nucleic acids, of which there are two closely related types: ribonucleic acid (RNA) and deoxyribonucleic acid (DNA). DNA is an elegant double-stranded molecule: it comprises two

Table 3.1 Overview of general terms of commonly occurring bases and ribonucleotides. Frequently used abbreviations are given in brackets. Note that the forms usually found in DNA (the 2'-deoxyribo forms) are referred to by prefixing with deoxy (or d for abbreviated names). Multiple phosphorylated nucleotides occur and abbreviated notations use D for diphosphate and T for triphosphate, e.g. ADP and ATP.

Base	Ribonucleotide (5'-monophosphate)
Purines (Pur)	
Adenine (Ade, A)	Adenylate (5'-AMP or Ado-5' or pA)
Guanine (Gua, G)	Guanylate (5'-GMP or Guo-5'-P or pG)
Pyrimidines (Pyr)	
Cytosine (Cyt, C)	Cytidylate (5'-CMP or Cyd-5'-P or pC)
Thymine (Thy, T)	Thymidylate (5'-TMP or Thd-5'-P or pT)
Uracil (Ura, U)	Uridylate (5'-UMP or Urd-5'-P or pU)

nucleic acid molecules that are complementary in sequence (as we describe shortly). This allows them to pair, forming the iconic double-helical DNA molecule we all recognize in Figure 3.1.

Cells have several types of RNA molecules: messenger RNA (mRNA) is a single-stranded copy of a gene sequence, made when it is needed and destroyed when no longer useful; transfer RNA (tRNA) and ribosomal RNA (rRNA) are stable molecules that are involved in the translation processes that convert gene sequence information copied into an mRNA molecule into proteins. Since the mid-1990s, a range of other RNA molecules have been discovered by scientists. These intriguing molecules have been shown to be important for the precise regulation of gene expression in all organisms. These regulatory RNAs include molecules referred to as small RNAs, micro RNAs (miRNAs), and long, non-coding RNAs. You can discover more about their role in another book in this series: *Genomics* by Julian Parkhill et al.

Nucleic acids have a directional polarity. This means the molecule increases in length in one direction only. This is all down to the bonding between the nucleotides. The 3'-OH of one nucleotide binds to the 5'-phosphate of the next using a phosphodiester linkage. As a result, one end of the polymer has a 'free' 5'-phosphate group, and the other end has a free 3'-OH group (see Figure 3.3). Nucleic acid sequences are written with the nucleotide containing the 5'-phosphate on the left, and move towards the 3'-OH group on the right; we say they are written in the 5' to 3' direction. The sequence of a polynucleotide is typically represented by using single capital letters to distinguish the different bases of the nucleotides (see Table 3.1).

Various atoms within nucleotides (and nucleosides) are able to interact with other atoms, particularly via the formation of hydrogen bonds (Figure 3.3). This means that the bases of some nucleotides are able to bind to certain others using hydrogen bonds. These interactions (called base pairing) allow different nucleic acid molecules to recognize each other. The favoured arrangements always involve purine–pyrimidine pairs. The bases that form pairs are said to be complementary. Generally, G pairs with C and A with T (in DNA) or U (in RNA). These different base pairs have similar overall shapes and are the basis of the double-stranded nature of DNA (see Case study 3.1). Scientists have used this inherent property to simplify how they describe DNA. Double-stranded DNAs are often written as one strand (usually 5′ to 3′) because the sequence of the other, complementary strand is understood as it is fixed by the base-pairing rules.

In Case study 3.1, we highlight that each DNA (and RNA) molecule has a unique three-dimensional structure that is specified by the sequence of bases in its chain. The way the bases stack together within double-stranded DNA dictates that it has to form a helix. Typically, DNA molecules form right-handed helices, and the most common conformation is the right-handed B-form (see Figure B in Case study 3.1). Various helical structures exist, and some of them are very similar to the standard B-form, such as the right-handed A-form. However, other structures are dramatically different, such as Z-DNA, which is a left-handed helix. Have a look at Figure 3.4 which shows you a 'normal' right-handed DNA helix beside a left-handed helix. Then have a look at lots of images of DNA on the internet and even in textbooks—it is amazing how often the helices are the wrong way round!

If you look carefully at Figure 3.3, you can see that the bases at the 5′-end of one strand pair with the bases at the 3′-end of the other. This is true for all double-stranded DNA molecules and the two strands are said to be 'antiparallel'.

As we have seen, the base pairing of nucleic acids contributes to their 'secondary structure'. Although most base pairing occurs between two separate strands, DNA can also form base pairs within a single strand (known as intra-strand pairing). This can lead to fascinating shapes forming within a nucleic acid strand, such as the hairpins shown in Figure 3.5(a). Single-stranded RNAs can also form intra-strand pairing, leading to the formation of many types of hairpin structures within them (Figure 3.5(c)).

Figure 3.4 A right-handed DNA double helix and an incorrect left-handed DNA double helix. See how many incorrect images you can find in books and on the internet!

Right-handed double helix   Left-handed double helix

Figure 3.5 Inverted repeat DNA sequences can adopt different types of structure. 'Inverted repeats' are repetitive DNA elements where the sequence is the same when the complementary strand is read in the same 5'–3' direction. (a) Such DNA sequences can exist in a regular double-stranded form. (b) Intra-strand base pairing allows the formation of a hairpin in a single strand. Due to their complementary sequences, both strands of DNA can take up a hairpin structure, which is referred to as a cruciform, or four-way junction. Intra-strand pairing can also occur in RNA molecules, as in the case of the tRNA shown here (c), where a region of double helix is highlighted in blue.

(a) Linear DNA

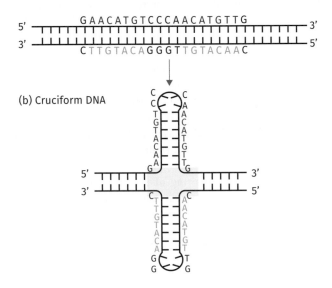

(b) Cruciform DNA

(c)

(a) and (b): Václav Brázda, Rob C. Laister, Eva B. Jagelská, and Cheryl Arrowsmith. Cruciform structures are a common DNA feature important for regulating biological processes. BMC Molecular Biology. © Brázda et al; licensee BioMed Central Ltd. 2011. https://doi.org/10.1186/1471-2199-12-33; http://creativecommons.org/licenses/by/2.0. (c): Adapted with permission from Crowe & Bradshaw, *Chemistry for the Biosciences*, third edition. Oxford: Oxford University Press, 2014.

Case study 3.1
The structure of the DNA double helix

DNA is the genetic material of all cellular organisms, and it provides the biochemical basis for the inheritable characteristics that are passed on to the next generation of cells. DNA was first discovered by Friedrich Miescher in 1869 when he was studying a compound he referred to as 'nuclein'. However, it was not until the middle of the twentieth century that DNA was universally accepted as the genetic material. By 1950, chemical analysis by another scientist, Erwin Chargaff, showed that DNA from all organisms had a 1:1 ratio of purine to pyrimidine bases. More importantly, the amount of dA was equal to dT and the amount of dG was equal to dC. This provided one of the first clues about the structure and function of DNA. These 'equivalence ratios', known as Chargaff's rules, gave rise to the idea of base pairing in DNA.

By this time, the chemical nature of DNA was starting to be understood, but the molecular details of its three-dimensional structure were still debated. This gave rise to one of the most well-known scientific controversies. Who deserves credit for establishing the structure of DNA?

The established narrative is that the concept of the iconic double-stranded helix, first published in 1953, was the work of Cambridge scientists, James Watson and Francis Crick (Figure A). But this doesn't tell the whole story. The data that Watson and Crick used to generate the molecular structure built on the work of Chargaff and his 'equivalence ratios'. But the vital evidence used by Watson and Crick were photographic images from X-ray diffraction studies that were provided by the physicist Maurice Wilkins, who was a lead scientist working in King's College London. The issue was that the image Wilkins supplied to Watson and Crick was not his own work. Rosalind Franklin was a research associate working with Wilkins. She was an extremely accomplished X-ray crystallographer who produced a key X-ray image of β DNA, now known as photo 51. This was the image that was offered to the two Cambridge scientists and it proved to be critical in allowing Watson and Crick to build and publish the first three-dimensional structure of this iconic molecule.

Now, there are many versions of this key event. In one version, Franklin was not consulted; Wilkins shared the image with Watson and Crick without her consent. In another, Franklin was already leaving Kings College because of 'scientific differences' with the Science Director at the time and Franklin was instructed to leave behind the data that she had produced—because it belonged to Kings. The implication was that nothing unusual had taken place when Franklin's photo 51 was shown to Watson and Crick: the photo belonged to the institute and not to the individual who had created it.

What there is little debate about is that Franklin received very little credit for her discovery until after her untimely death in 1958—she was only 37 when she died of ovarian cancer. Instead, Crick, Watson, and Wilkins shared the Nobel prize for Physiology or Medicine in 1962 for the identification of this elegant structure. The Nobel prize can only be shared by a maximum of three people and Franklin was not part of the Nobel prize-winning team. This is often used as an example of the way her contribution was minimized, which may well be true—but another factor is undoubtedly that the prize

Figure A Although James Watson and Francis Crick are the names most commonly associated with the structure of DNA, the work of Rosalind Franklin and Maurice Wilkins played an equally important role in the ground-breaking discovery

Watson   Crick   Wilkins   Franklin

World History Archive/Alamy Stock Photo; Science History Images/Alamy Stock Photo; Archive PL/Alamy Stock Photo; Pictorial Press Ltd/Alamy Stock Photo

cannot be awarded posthumously, and in 1962 Rosalind Franklin had sadly been dead for four years.

Identification of the structure of DNA led to the 'age of molecular biology', and this has been at the forefront of basic biological research ever since.

An important aspect of the Watson and Crick model is that two single-stranded DNA polymers come together to form double-stranded helical structures (Figure B). The primary structure of each single strand (determined by the order of bases that are joined together to form it) dictates how they will interact because hydrogen bonds form between bases on the two opposite strands (Figure 3.3). Base pairs between dG·dC and dA·dT are central to the molecular model proposed in 1953 and they are frequently referred to as 'Watson–Crick' base pairs. It is these base pairs that 'lock' the structure of regular double-helical DNA. In these base pairs, three hydrogen bonds are formed between dG and dC and two hydrogen bonds are formed between dA and dT. This means that dG·dC bases are able to interact at higher temperatures than dA·dT pairs—they are said to have greater thermodynamic stability.

An important characteristic of the two Watson–Crick base pairs is their similar overall shape (or 'isomorphism'). Their similar chemical structures allow them to be replaced by each other without disrupting the backbone structure of the double helix. DNA that consists completely of Watson–Crick base pairs can form a regular double stranded or **duplex** structure with the same overall conformation throughout the duplex, independent of the particular sequence of bases—this is the 'B-form' helix shown in Figure B.

Watson and Crick realized that the base sequence of one strand in double-stranded DNA dictates the sequence of the second complementary strand. It is this property of DNA that is fundamental to its ability to store information in biological form and for this information to be efficiently copied into new DNA molecules, through the process of replication. Base pairing rules are also important for directing the synthesis of mRNA molecules, which are produced during the transcription of specific DNA sequences.

Figure B These images show two different representations of the structure of double-stranded DNA, as identified by Crick, Franklin, Watson, and Wilkins. This is the structure for the typical, or average, type of DNA molecule, and is named B-DNA. (a) Cartoon representation highlighting some of the important parameters and distances in one turn of double-stranded DNA. Note that the distance between base pairs is approximately 0.34 nm (3.4 Angstrom, Å). (b) Molecular model of the structure shown from the side (left image) and looking down the helix (right image).

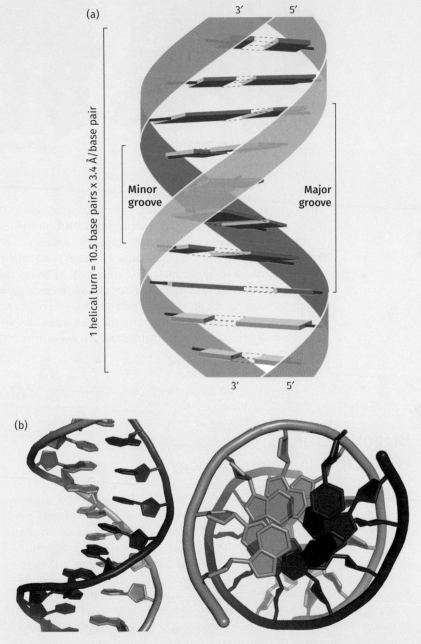

(a): Adapted with permission from Craig, et al. *Molecular Biology Principles of Genome Function*, Oxford: Oxford University Press. Copyright © 2014. Reproduced with permission of the Licensor through PLSclear. (b): Reproduced with permission from Crowe & Bradshaw, *Chemistry for the Biosciences*, third edition. Oxford: Oxford University Press, 2014.

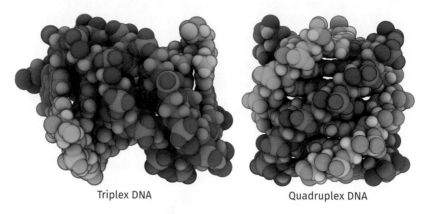

Figure 3.6 Three-dimensional space-filling models for triplexes (left) and quadruplexes (right). Atoms shown in different colours are contained within the different strands in the DNA molecules.

Triplex DNA   Quadruplex DNA

Laguna Design/Science Photo Library.

Under certain conditions, unusual base pairs can occur. These allow a higher number of strands to interact to form DNA molecules. These types of structures are favoured by specific types of sequences and can include three-stranded molecules, referred to as 'triplexes', and four-stranded molecules, referred to as 'quadruplexes' (Figure 3.6). Although these structures appear to be unusual in average DNA sequences, there is growing evidence that they can occur within cells and that they may have physiological functions. Some of these unusual structures for DNA molecules are also important for nanotechnological applications (see Bigger picture 3.1).

Bigger Picture 3.1
Using DNA in nanotechnology

Naturally occurring nucleic acids perform many biological functions, from information storage to the catalysis of biochemical reactions. We have seen that they can also adopt a wide range of different structures, which are stabilized by specific sequences within the nucleic acids. Recently, biochemists and molecular biologists who study the links between sequence and structure have begun to plan ways to use DNA molecules in different types of nanotechnological applications. Studying DNA in this way has become known as 'DNA origami': it describes ways in which long strands of DNA can be designed so that they fold into specific two-dimensional or three-dimensional shapes. Research is focusing on the concept that, once folded into these complex and stable shapes, the DNA could deliver drugs to where they are needed in the body, or release compounds to the environment at specific times (Figure A).

Figure A Examples of applications of DNA origami in different areas of science. Sequence and structural properties of DNA molecules are being developed for a wide range of nanotechnological applications, including: (i) biomaterial science; (ii) nanomedicine; (iii) nanorobotics; and (iv) molecular computation.

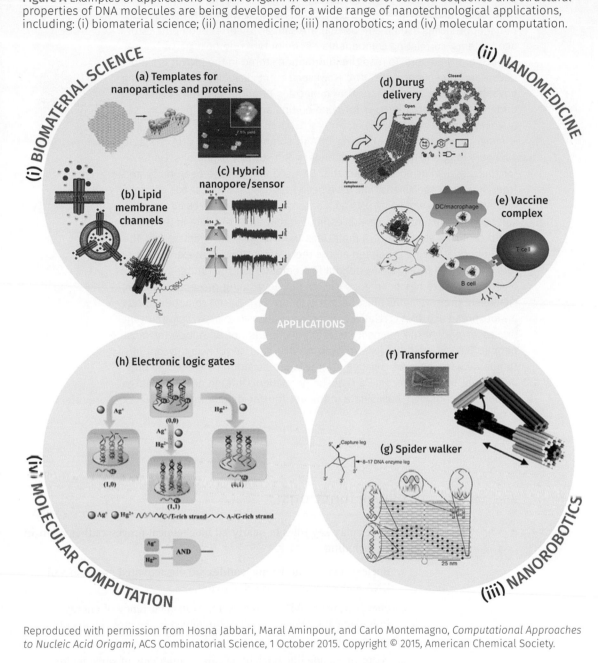

Since natural DNA is remarkably efficient at storing genetic information, scientists have begun to consider whether it could be used as the storage medium in computers. Research in this area has progressed significantly and some successes have been reported. The intrinsic stability of the DNA polymer—with its phosphate backbone and its stable hydrogen base pairs— strongly suggests that DNA storage systems could have significantly longer shelf lives than those of current storage disk drives. Different approaches have been taken to store and decipher information within DNA sequences. Initial approaches used methods that are similar to current computer storage systems, correlating components of computer code or logic with a sequence of DNA letters. To read the information stored in the DNA, it is simply necessary to sequence the DNA. However, recent developments are taking advantage of the potential for DNA molecules to adopt different structures under different conditions, or for gene editing approaches to rearrange sequences in controlled ways.

A significant advantage of using DNA in computers is that DNA is stable over very long timeframes, at least under good storage conditions. Evidence for this stability includes the fact that DNA sequences have been recovered from biological specimens that have been fossilized for thousands of years. Another advantage of using DNA as a data store is that complex information can be stored in a tiny space. It is estimated that all of the data in the entire world could be stored in DNA molecules that weigh a few grams. Since the amount of data being generated across the world is growing at extraordinary rates, this could be extremely helpful for computers of the future!

Although there has been good success in developing these approaches, significant improvements must be made before DNA-based computing finds a widespread use. DNA is relatively slow and expensive to make using current processes and, at the moment, mass production is not an option for most people. But, as methods for synthesizing DNA and sequencing it continue to improve and become cheaper, scientists are becoming increasingly optimistic that DNA will become a realistic permanent storage device for large amounts of computer-based data in the not-too-far-distant future.

Cellular functions

Nucleotides play key roles in nearly all biochemical processes that occur in all cells, including:

1. nucleic acid synthesis: nucleotides are the activated precursors of DNA and RNA, as discussed above;

2. energy transfer: ATP is used as a universal currency of energy in biological systems and GTP is involved in the movements of macromolecules;

3. coenzymes: adenine nucleotides are components of some major coenzymes;

4. metabolism: nucleotides can be converted to new molecules that are involved in many biosynthetic reactions, and help to regulate general metabolism.

In the rest of this chapter, we will focus on general cellular functions associated with nucleotides. We will discuss points 2–4 in more detail in Chapter 4.

The quantity of nucleotides in different cells is linked to the rate of their growth, meaning that there can be striking variations between different cell types. It is estimated that nucleotides and nucleic acids account for 5–10 per cent of the total weight of rapidly dividing bacteria. In mammals, the amounts vary from one tissue type to another, from approximately 1 per cent in muscle to 15–40 per cent in sperm cells.

There are always lots of free nucleotides in the cytoplasm and other parts of the cell, and they are in constant flux between their free and polymeric states. However, the movement from the free to the polymerized forms has to be tightly controlled within cells. This is illustrated most vividly with DNA—an inadequate supply of any of the free nucleotides is lethal, but an overabundance of free nucleotides can lead to an increase in the number of mutations in the genome. Cells rely on small pools of free nucleotides (with the exception of ATP), which they keep topped up by constantly synthesizing new nucleotides, especially during times of high need, such as replication of DNA during cell division.

In addition to the major nucleotides listed in Tables 3.1 and 3.2, many other types are present in small amounts within cells. Some of the minor nucleotides are required for specific cellular functions and are particularly common in rRNA and tRNA. In some cases, they are the unwanted by-products of cellular reactions: they can be damaging to cells and affect their survival. For example, DNA damage, perhaps caused by X-rays or UV radiation, can generate a variety of unusual nucleotides. Fortunately,

Table 3.2 Amounts of nucleotides in bacterial and mammalian cells. Values shown for bacterial cells are for *Escherichia coli* and *Salmonella typhimurium* in exponential growth, and those for mammalian cells are determined from diverse mammalian tissues. The terms dNMP and NMP refer to nucleotides containing all bases.

Nucleotide name	Amount in bacterial cells (mmol l^{-1})	Amount in mammalian cells (mmol L^{-1})
DNA precursors		
dNMPs	43	n.d.
dATP	0.18	0.013
dGTP	0.12	0.005
dCTP	0.07	0.022
dTTP	0.08	0.023
RNA precursors		
rNMPs	270	n.d.
ATP	3.0	2.8
GTP	0.92	0.48
CTP	0.52	0.21
TTP	0.89	0.48

Note: n.d. = not determined.

Kornberg, A. and Baker, T., *DNA Replication*, second edition. New York: W. H. Freeman, 1992, 54.

a wide range of metabolic pathways have evolved in cells that allow their removal, so they do not cause errors in cell division or protein synthesis.

Evolution has also enabled cells to use nucleotides with more complicated structures than those shown in Figure 3.3. Variants of purine nucleotides exist that have additional polyphosphates (e.g. ppGpp and pppGpp). Other alternatives include cyclical molecules (e.g. cyclic AMP (cAMP) and cyclic GMP). These molecules often play essential signalling roles within the cells that synthesize them or between neighbouring cells. For example, cAMP (Figure 3.7(a)) forms inside a cell when the cell receives a signal in the form of a peptide hormone, such as adrenaline, glucagon, or follicle stimulating hormone, which locks on to receptors located on its surface. Its formation can then be used to trigger some kind of response in the cell. Cyclic AMP is also a key chemical that allows simple organisms such as slime moulds to communicate, where the cAMP provides the signal for individual cells to come together to form a reproductive body (Figure 3.7(b)).

The majority of pyrimidine nucleotides in cells exist as DNA and RNA. For example, cytosine is found in both DNA and RNA. This is not the case with uracil and thymine: uracil is only present within RNA and thymine is, in general, only found in DNA. Because of the pairing of specific bases (T with A, C with G), pyrimidine nucleotides will always be 50 per cent of double-stranded DNA. The ratio of cytosines to thymines is fairly constant within a single species, but this ratio does vary between species. This situation is particularly noticeable in species that live in extreme environments, such as high temperatures, where a high frequency of C•G base pairs helps increase the stability of the DNA molecule by making them more heat-resistant.

Nucleotides containing adenine and guanine are used in many aspects of cellular biochemistry. As with the pyrimidine nucleotides, a large proportion of the major purine nucleotides are found in polymers of both DNA and RNA. However, significant amounts of certain purine nucleotides play a part in a wide variety of energy-requiring processes.

Figure 3.7 Cell signalling events are promoted by cyclic AMP (cAMP). (a) Chemical structure of cAMP; (b) cAMP is the signal that makes hundreds of thousands of cells come together to form the body of a slime mould like this one.

(a)   (b)

(b): Anthony Short.

All life forms require modified versions of purine nucleotides, as previously highlighted with cAMP. Reduced and oxidized forms of nicotinamide adenine dinucleotide (NADH and NAD$^+$, respectively) and different forms of flavin adenine dinucleotide (FAD, FADH, FADH$_2$) are also used in a variety of oxidation–reduction reactions that occur within cells. The concentrations of these modified nucleotides can vary significantly; some, such as cAMP, are present for a relatively short time because they are degraded rapidly by enzymes when no longer needed.

Biochemists have also synthesized many novel modified nucleotides, which have found many uses in *in vitro* and *in vivo* experiments. For example, fluorescent derivatives of nucleotides are very important for studying the role of nucleotides and nucleic acids, as they allow these molecules to be seen within cells. Furthermore, such molecules are used during high-throughput automated sequencing of DNA, which has dramatically impacted on the amount of data that is generated and the way it is used (see Bigger picture 3.2).

In addition to their distinct roles in DNA and RNA, the most important property of purine-based nucleotides is their role as an energy pool for many enzymatic processes. We highlighted the ability of ATP to act as a universal energy currency at the start of this chapter and will discuss it further in Chapter 4. GTP has similar properties for some specific processes, such as formation of proteins and signalling pathways.

Chromosomes and genomes

Each somatic cell in an organism has the same amount (and sequence) of DNA, but these amounts vary dramatically between species. An organism's genome is organized into a specific number of DNA–protein complexes, usually called chromosomes. The smallest genomes are carried by viruses and may contain fewer than 10^4 base pairs. The bacterium *Escherichia coli* has one chromosome of about 4.5 × 10^6 base pairs and the unicellular eukaryote *Saccharomyces cerevisiae* has sixteen chromosomes comprising a total of approximately 2 × 10^7 base pairs. The DNA content of humans is approximately 3 × 10^9 base pairs in total, carried on twenty-three different pairs of chromosomes (forty-six chromosomes in each somatic cell). Some organisms have even larger genomes, with a rare Japanese flower named *Paris japonica* reported to contain 149 billion base pairs—fifty times as much DNA as humans!

Natural DNAs are among the longest molecules in the world. If the longest human DNA molecule (chromosome 1) was stretched out in its double-stranded form, it would be almost 8 cm in length. Taking into account all chromosomes, each human cell contains almost 2 metres of double-stranded DNA! This is almost 20,000 times larger than the diameter of a typical human cell (100 μm), indicating that DNA molecules have to be very flexible to fit and fold into such tiny spaces. This impressive feat is achieved by the binding of the DNA to several different proteins, which help it to coil into the tightly-packed chromosomes. These folded protein–DNA complexes are approximately 40,000 times shorter than a double-stranded DNA strand that is pulled out taut. Each chromosome consists of two strands of DNA, known as chromatids, which are joined by a centromere.

Despite being so tightly packed, the sequence information carried by DNA is not corrupted and it is still accessible for the vital cellular process of DNA replication and transcription. Among all of these facts and figures, think about this: since a human contains almost 40 trillion cells, if the DNA from every cell in your body was stretched out in a line, it would reach the Sun and back more than 250 times! (It would also be impossible to see with the human eye because, at 2 nm (Figure A in Case study 3.2), the width of a DNA molecule is 50 000 times narrower than the average human hair.)

Significant advances in DNA sequencing technologies since the early 2000s have allowed us to identify full genome sequences for many organisms (see Bigger picture 3.2). Detailed analysis of this data has revealed a surprising fact. Much of the DNA in most organisms does not encode protein sequences: it is non-coding DNA. So what does this non-coding DNA do and why do genomes have it? It appears that some of these sequences are transcribed into important RNA molecules (e.g. tRNA, rRNA, and regulatory RNAs). Non-coding DNA sequences also regulate transcription and translation of protein-coding sequences and the replication of DNA. It also forms specific physical structures within chromosomes, such as centromeres and telomeres.

The proportion of coding DNA and non-coding DNA within genomes varies greatly between organisms. Only 2 per cent or so of the human genome codes for proteins, while more than 80 per cent of a typical prokaryote genome is coding DNA. The complexity of an organism is not always related to the size of its genome, and much of the unexpected genome size of some organisms appears to be due to variations in the extent of non-coding DNA. This suggests that a lot of biological complexity stems not from the number of genes in an organism but from the intricate way in which those genes are regulated. If we can assume a consistent proportion of non-coding DNA is regulatory, the more non-coding DNA there is, the greater the regulation of gene expression that is possible.

Although DNA tends to get the glory as the genetic code molecule, the role of RNA should not be underrated or forgotten. The majority of cellular mechanisms that coordinate the expression of genes and their encoded proteins (turning them on and off at specific times in the cell life cycle or in response to environmental signals) require RNA molecules to operate. Their small size and their ability to bind to other nucleic acids mean that RNA molecules are extremely well suited to these roles. Different types of RNA vary in size—from approximately 20 nucleotides for miRNAs, 65–75 nucleotides for tRNAs, and up to several thousand bases for mRNAs. As already highlighted, intra-strand base pairing within RNA molecules leads to three-dimensional structures that are specific to each type of RNA and their individual roles. However, all double-stranded regions of RNA have a helical structure that resembles the right-handed A-form helix of DNA.

The specific shapes taken up by some RNAs give them biochemical properties that were very surprising when initially discovered: they are able to catalyse specific chemical transformations—in other words, they act as enzymes! Such catalytic RNAs are often referred to as ribozymes, and they undertake important biochemical roles in all cells. For example, the cellular organelles called ribosomes that are involved in protein synthesis use catalytic RNAs to help link amino acids together. Biochemists have recently discovered other reactions that these molecules can do, and there is growing belief that they were the first types of biological catalysts to exist on Earth. Eventually, proteins took over as the dominant biological catalysts on Earth because they are able to perform a wider range of reactions at faster rates. However, the identification of catalytic RNAs has brought forward interesting ideas about the origin of terrestrial life.

The 'era of molecular biology'

As biochemistry developed as a distinct area of science in the early part of the twentieth century, it became clear that individual molecules were critical for biological processes. The concept of 'molecular biology' began to be established in the 1930s, but it was the solving of the structure of DNA that led to 'the era of molecular biology'. Once we understand the structures of DNAs and RNAs, we can get an insight into how their biological functions link with other molecules, particularly proteins. And our increasing knowledge of these sophisticated molecules is leading to improved technologies, enabling us to manipulate the molecules in biochemical experiments.

Technological developments in molecular biology have led to particular advances when it comes to the manipulation of nucleic acids. During the 1960s, a range of enzymes was identified that can cut DNA molecules. These so-called restriction enzymes recognize specific sequences within DNA molecules and break them in reliably consistent ways; this controlled cutting-up of DNA is used by scientists as a laboratory tool. Scientists have also identified and purified a completely different set of enzymes, DNA ligases, that are able to join breaks in DNA molecules if correct base pairing is in place. By combining the biochemical activities of restriction enzymes and DNA ligases, molecular biologists developed DNA cloning. Gene sequences can be cut out and moved between different DNA molecules, or even different organisms. These novel DNA molecules, which are not present in nature, are called recombinant DNAs.

Many technical developments have advanced and improved the experimental approaches used by molecular biologists. One of the most important leap forwards was the development of the polymerase chain reaction (PCR). This now has several different formats, but the most widespread, standard format was developed by Kary Mullis, a maverick American biochemist who won the 1993 Nobel Prize in Chemistry for this work, using enzymes from bacteria originally found in the hot springs of Yellowstone Park.

PCR is an *in vitro* technique that allows the amount of a specific DNA sequence to be rapidly increased (or 'amplified') (Figure 3.8). The method developed by Mullis used repeated cycling of DNA to high temperatures, allowing the nucleotide strands to separate (a process known as denaturing). Once the sample cools to lower temperatures a small oligonucleotide binds to the DNA sequence of interest. This can act as a 'primer' to allow new DNA synthesis to be carried out by the enzyme DNA polymerase.

Figure 3.8 Large-scale amplification of amounts of specific DNA using the polymerase chain reaction (PCR)

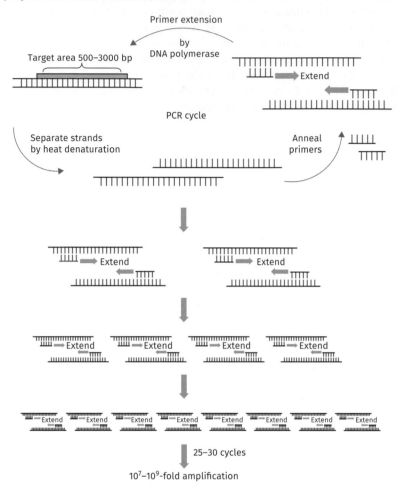

The speed of PCR was enhanced by introducing the use of *Taq* DNA polymerase (an enzyme used for DNA synthesis by an organism that grows at very high temperatures) because this allowed repeated thermal cycles to be conducted in an automated way (whereas 'standard' DNA polymerase would cease to operate at such high temperatures). A variety of PCR approaches have now been developed and these have been applied extensively to the identification, detection, and diagnosis of genetic sequences and infectious disease. For example, PCR is the cornerstone of current forensic techniques that are used to identify criminals.

As we look back across the era of molecular biology, it is clear that this variety of enzyme gave scientists the opportunity to make dramatic changes to the genetic make-up of organisms. However, it is important to remember that these workhorse laboratory enzymes have important biological roles in their native cells. For example, restriction enzymes are critical to the way that bacterial cells protect themselves from hostile viruses; they are part of their 'immune' system that attacks foreign viral DNA. Similarly, DNA ligases are essential for all cells because they participate in the final step of all reactions that involve DNA synthesis, including replication.

These experimental tools, such as restriction enzymes and PCR, have helped scientists to improve their understanding of the biochemistry of cells. This knowledge is now being used to develop new drugs and more efficient diagnosis protocols for human diseases. As molecular biology tools have become better, faster, and cheaper, the potential for us to genetically engineer or modify organisms has improved significantly. Such experimental approaches are now widely exploited in biotechnology industries and are beginning to impact on therapies for human diseases, especially through gene therapy. We will look at some relevant examples of these applications in Chapter 6.

From gene synthesis to synthetic biology

Biochemists have helped identify the chemical structures of a wide range of nucleotides, and many can now be synthesized in the laboratory. These compounds have been useful in different ways, including as potential pharmaceuticals, and they are particularly valuable in the scientific analysis of the metabolism of nucleotides and nucleic acids. We can now synthesize large amounts of nucleotides and, in recent years, the quality of these molecules has improved significantly and their cost has reduced dramatically. The improved availability of fluorescent derivatives of nucleotides has been fundamental to the development of high-throughput automated sequencing of DNA (see Bigger picture 3.2).

The availability of large amounts of nucleotides at low cost has also led to technologies that allow large DNA molecules to be synthesized directly *in vitro* (see Chapter 7). These technologies allow dramatic and ambitious changes to be incorporated into DNA sequences. Scientists have even begun to develop completely synthetic genomes, including studies that are exploring what the minimal genome for a fully function microorganism might be. Craig Venter is a controversial biochemist and biotechnologist who was also at the forefront of studies to sequence the human genome (see Bigger picture 3.2). By 2010, the work being undertaken by Craig Venter and his colleagues culminated in the generation of a living organism with a fully synthetic genome that was capable of replicating itself billions of times. The newly created organism had a complicated scientific name (*Mycoplasma mycoides* JCVI-syn1.0 or *Mycoplasma laboratorium*), but is more commonly referred to as 'Synthia'!

Improvements in technologies that manipulate and analyse DNA have led to the emergence of a new area of science called synthetic biology, which has developed rapidly in the twenty-first century. This discipline uses engineering principles to take advantage of and improve biological systems, usually with applications to tackle specific problems. Synthetic biology research often follows methods used by engineers to 'design, build, and test' and then repeats this cycle until the system works as required. The 'biology' aspects add significant challenges to problems usually tackled by engineering, but solutions to some important challenges are gradually being identified. Synthetic biology has been proposed as one of the scientific technologies that will help address the challenges in health, energy, and food security that societies across the globe will face in the twenty-first century. We will review examples of some of these areas in later chapters, and synthetic biology is discussed more widely in other titles in this series.

As we will see in the later discussions, potential applications of synthetic biology and genetic engineering to situations outside of laboratories bring significant ethical, legal, and regulatory dilemmas. These are particularly challenging if they require the development of genetically modified organisms, which have seen slow progress due to the distrust that has developed with some wider populations of society (see Chapter 7). These developments have demonstrated that it is important to ensure good engagement between scientists, funders of the research, and the wider general public. This has led to increased visibility for ethical considerations of synthetic biology and genetic engineering research, particularly in relation to its potential impact on society and the environment.

Bigger picture 3.2
The impact of DNA sequencing technologies

We now take it for granted that DNA can be sequenced relatively quickly and cheaply, and this is indispensable for basic biochemical and biological research. It is also transforming personalized medicine, biotechnology, forensics, and **bioinformatics**. But sequencing of DNA has not always been straightforward. In fact, it was not until the 1970s that the first methods of DNA sequencing became available to laboratories using state-of-the-art molecular biology methods and equipment. Several approaches were initially tested, with two very different methods becoming popular: one used chemicals to attack specific bases and the other took advantage of DNA polymerases, enzymes that are normally involved in the synthesis of DNA. Both methods resulted in the Nobel Prize in Chemistry in 1980, but, within a few years, the enzymatic methods became the 'go to' choice of almost all nucleotide sequencing projects.

The earliest forms of nucleic acid sequencing were developed for both RNA and DNA, and allowed the complete genome of viruses made from each type of nucleic acid to be sequenced. Sequencing of RNAs required the creation of a DNA copy using a specific type of DNA polymerase and short sequences of nucleotide primers to allow DNA synthesis to start. This technique was the forerunner to similar methods with DNA, with various types of DNA polymerase being tested for catalysis of the new DNA. The British biochemist Frederick Sanger improved the DNA sequencing methodologies, and he was one of the scientists awarded the Nobel Prize in 1980. (This was Sanger's second Nobel Prize in Chemistry: he had previously been awarded one in 1958 for his biochemical studies on the structure of proteins, especially insulin. He is one of only four people to win two Nobel Prizes, and one of only two people to have done so in the same category.)

The early sequencing methods relied on radioactive nucleotides. The safety procedures for working with radioactivity and the methods used were

Figure A Fluorescent nucleotides increase the throughput of DNA sequencing. (a) Raw sequence data for automated enzymatic sequencing of DNA. The four colours indicate the relative position of the bases in the DNA fragment. Each four-colour vertical line corresponds to a different sequence reaction. Samples that are nearer the cathode (towards the bottom of the image) are smaller and these bases are closer to the primer that initiates the reaction. (b) Portions of a representative analysed sequence determined by the automated sequencer.

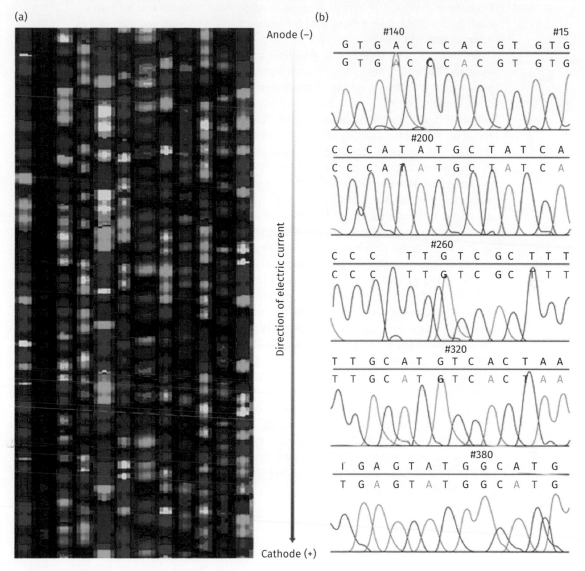

laborious and time-consuming. Major steps forward in DNA sequencing occurred with the production of large amounts of relatively cheap (non-radioactive) fluorescent nucleotides with different colours used for each of the four bases. The different colours of each of the four bases can be differentiated in a single reaction (Figure A). This improved the throughput so DNA sequencing could be easily automated, leading to dramatic improvements in the speed and accuracy of these approaches.

For a few years the methods used for sequencing improved gradually, but during the 1980s a sea change occurred, because the processes became automated. Vast improvements in computing technologies led to sequencing of complete genomes, with the first bacterial genomes being completed from the mid-1990s. The human genome sequence was completed in its first draft form in 2001, taking more than ten years of scientific work and costing approximately $3 billion.

Around this time, several new methods for DNA sequencing were developed and these have been called the 'next-generation' or 'second-generation' sequencing methods. A variety of processes is involved in these methods, some building on the use of DNA polymerases, especially in combination with PCR to amplify the amounts of DNA. However, some recent techniques focus on different properties of nucleotides and nucleic acids, such as the detection of hydrogen ions that are released during DNA synthesis, or the varying electrical or ionic properties of the different nucleotides.

Further advances in sequencing methods are being developed by the incorporation of mass spectrometry or microscopy-based techniques, such as atomic force microscopy or transmission electron microscopy. All of the newer sequencing methods are characterized by being highly scalable, allowing much larger DNA molecules to be sequenced. It is even possible to consider sequencing entire genomes because multiple fragments are processed at the same time, giving it the name 'massively parallel sequencing'.

The many different types of sequencing technologies have dramatically improved the speed of data recovery, reduced cost, or both. This means it is now possible to sequence the complete genomes of organisms in just a few days for a few thousand pounds. We already have access to genome sequences for many thousands of different organisms, but future developments will make it possible to sequence the genome of any person at costs and timescales that are affordable for healthcare systems. This really will open the door to personalized medicine, where the genes of individuals can be identified to assess the chances of drugs being useful for specific individuals. These sequencing technologies are offering huge opportunities for biological and medical sciences, but they also bring many challenges in understanding how to analyse such large quantities of data (see the Genomics title in this series).

Chapter summary

- Nucleotides are essential for the correct functioning of all cell types because they play a central role in a wide variety of cellular processes.
- Cells contain many types of nucleotides, with each one consisting of a nitrogen-containing base, a five-carbon sugar, and one or more phosphate groups. Usually, differences between nucleotides occur within the base portion of the molecule.

- The best known and biologically studied nucleotide is adenosine-5′-triphosphate, usually referred to as ATP. It has several important functions in cells, but its most important property is to transfer energy in the reactions catalysed by many enzymes involved in metabolism.
- Common bases are referred to as purines, consisting of adenine and guanine, and pyrimidines, consisting of cytosine, thymine, and uracil. Two bases that are said to be complementary can interact to form base pairs, with adenine able to pair with thymine (or uracil) and guanine able to pair with cytosine.
- Nucleotides are critical for the storage and usage of genetic information in their polymeric forms, known as nucleic acids. Nucleic acids exist in two types, ribonucleic acid (RNA) and deoxyribonucleic acid (DNA).
- DNA molecules are usually formed from two polymers that contain complementary sequences, allowing formation of stable pairing to form the basis of the double-stranded nature of DNA as the genetic material.
- RNA molecules are present in cells as different types of single polymeric chains. Different forms of these molecules are messenger RNA (mRNA), transfer RNA (tRNA), ribosomal RNA (rRNA), and micro RNAs (miRNA), which are all involved in processes that convert or regulate how gene sequence information is transcribed and translated into proteins.
- The deep understanding of DNA and RNA molecules and the cellular processes that they are involved in underpins the science of 'molecular biology'. This knowledge led to the development of techniques that manipulate genes, allowing their cloning into novel forms, known as genetic engineering (or modification). These approaches are fundamental to biotechnology, which allows biological molecules to be applied to solve major challenges. More recently, these methods have developed into synthetic biology, which aims to overcome challenges to society by combining engineering principles with biology.

 ## Further reading

Carey, N. (2018) *Junk DNA: A Journey through the Dark Matter of the Genome*. Columbia University Press, New York.

Maddox, B. (2003) *Rosalind Franklin: The Dark Lady of DNA*. HarperCollins, New York.

Ridley, M. (2004) *Nature via Nurture: Genes, Experience and What Makes Us Human*. HarperPerennial, New York.

 Discussion questions

3.1 What proportion of the human genome codes for proteins? The remaining DNA sequences that do not code for proteins are sometimes referred to as 'junk DNA'. What biological functions can these sequences have within human cells?

3.2 In addition to the common nucleotides containing adenine, guanine, cytosine, and thymine, DNA molecules often contain unusual types of nucleotides. What unusual nucleotides are found and how are these formed? What biological effects may occur due to the presence of these nucleotides?

3.3 At a chemical level, RNA molecules are more unstable than DNA. Why?

3.4 Advances in sequencing methods will soon make it affordable to sequence the entire genome of every human, which will open the door to personalized medicine. Suggest potential benefits and problems that might result. What kinds of ethical and moral dilemmas will this opportunity bring?

3.5 Some RNA molecules can act as enzymes. What kinds of reactions are catalysed by these molecules? Why do you think these types of catalytic RNAs are likely to have been critical for the evolution of life on Earth.

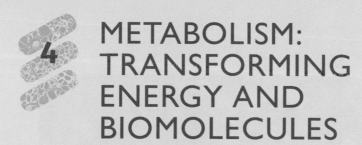

4 METABOLISM: TRANSFORMING ENERGY AND BIOMOLECULES

Every cell has a set of biomolecules that are responsible for its biological functions. These biomolecules are constantly being synthesized, degraded, and converted to new molecules. This dynamic process is known as **metabolism**. Although there are many different sets of reactions, or **metabolic pathways**, only a small number are fundamental to all cells. Here we will review two metabolic pathways that are particularly well understood by biochemists because of their importance for life. These pathways are aerobic **respiration** and **photosynthesis** (Figure 4.1).

To appreciate cellular metabolism, it is important to understand how energy is transferred between different molecules and converted to different forms. Moving and transforming energy is fundamental to how cells function, and the biochemists that study this topic describe it as **bioenergetics**. One of the key theoretical concepts that is particularly useful for understanding this biochemical topic is '**free energy**'. This concept helps identify whether biochemical reactions are favourable, or not—that is, how likely they are to happen under specific cellular conditions. We will start our study of metabolism and bioenergetics by looking at adenosine triphosphate (ATP), a molecule that is fundamental for all cells through its role as an energy store that links favourable reactions with those that would not, or could not, normally take place.

Figure 4.1 Animals such as this rufous-tailed jacamar in the Caribbean, along with the plants around it, all depend on aerobic respiration and photosynthesis to survive—but only the plants carry out photosynthesis

Anthony Short.

Transforming energy in cells

In Chapters 1–3, we have mentioned some of the chemical reactions that synthesize and degrade the compounds that are critical for cells to function. Since all of these reactions take place inside cells, they often influence each other. In fact, the products of one reaction often form the substrates of a different reaction. The sum of all processes that take place within each cell is referred to as its metabolism. This includes all of the series of reactions that lead to the synthesis, degradation, or transformation of specific compounds, usually referred to as metabolic pathways. In previous chapters, we have mentioned how these processes provide the different biomolecules that are the basic components of cells. But how do these critical reactions and metabolic pathways satisfy the energy requirements of cells?

We will answer this question using aerobic respiration and photosynthesis as examples. In so doing, we will describe energy transduction: the way energy is transferred between different molecules. Much of the information we have about these biochemical processes comes from laboratory experiments that provided plenty of data. However, it was an appreciation of thermodynamics that allowed biochemists to interpret these results and to fully understand how, why, when, and where these intricate biochemical pathways take place in cells (see Scientific approach 4.1).

Scientific approach 4.1
When energy is 'free' but costly

The laws of thermodynamics apply to **all** systems, including those in living organisms. The terminology used in thermodynamics often has crossover with everyday words, but with very different meanings! So, to avoid confusion, in thermodynamics energy is the capacity for any system to do work—the process of acting against an opposing force. **Thermodynamic parameters** refer to a system and its surroundings. Everything that is not a part of the system is known as the surroundings. These parameters must have a clear boundary, which can be defined in different ways by biochemists. For example, the system could be an atom or group of atoms, or molecules, cells, tissues, or an entire organism, depending on the experiment being performed.

The **total energy** of a system is the sum of both its **kinetic** and **potential energies**—the energy associated with movement and of stored energy, respectively. Provided no external forces act on a system, its total energy is constant.

There are three main laws of thermodynamics, but just the first two are most relevant here:

- The **First Law of Thermodynamics** (sometimes referred to as the Law of Energy Conservation): '*In any physical or chemical change the total amount of energy in the system remains constant, although the form of the energy may change.*' You may be more familiar with this expressed as '*Energy cannot be created or destroyed, just converted from one form to another.*' Two factors are fundamental to this topic: **enthalpy** (or the energy transfer that occurs during a reaction, represented by the symbol H) and **entropy** (the extent of randomness or disorder, represented by the symbol S).

- The **Second Law of Thermodynamics**: '*The tendency in Nature is towards maximum disorder. Therefore, the total entropy of a system (and its surroundings) always increases for a spontaneous process.*' For example, in living systems, once an organism dies no energy is used to keep it ordered and the organism starts to decay (see Figure 4.2): its matter becomes disordered.

'Free energy' has been defined in several ways over the years, but it is generally understood as a measurement for the ability of a thermodynamic system to use its energy to cause change. In other words, it is the energy that is 'available to do useful work'. As long ago as the 1870s, this concept became popularized by the American scientist Josiah Willard Gibbs (Figure B). In fact, he even had the concept named after him: Gibbs free energy (also represented by the symbol G).

In simple thermodynamics terms, Gibbs free energy is the maximum amount of reversible work that may be performed by the system if it is kept at a constant temperature and pressure. Now, things are complicated somewhat because calculations need to take into account changes in entropy (the urge to become disordered) in both the system *and* the surroundings. Gibbs discovered how to address this issue. He devised a formula that combined the first and second laws of thermodynamics, and he demonstrated that, during chemical reactions, G can be described by three thermodynamic quantities:

1. The change in enthalpy (ΔH), reflecting the change in the number and kind of bonds in substrates/products.
2. The change in entropy (ΔS), reflecting the change in order during the reaction.
3. Temperature (T), the absolute temperature (measured on the Kelvin scale; 273 K = 0°C).

Figure A shows this relationship written as a relationship using mathematical notation.

What Gibbs formulated explained how thermodynamic laws can be applied to biological systems. To put it simply, a reduction in free energy, or G, is absolutely required for biological processes to take place spontaneously at constant pressure and temperature. When reactions take place, then there is a general natural tendency to move towards a minimum of the Gibbs free energy.

Figure A This formula can be used to apply thermodynamic laws to biological systems

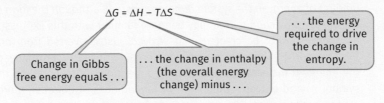

$$\Delta G = \Delta H - T\Delta S$$

... the energy required to drive the change in entropy.

Change in Gibbs free energy equals ...

... the change in enthalpy (the overall energy change) minus ...

Figure B Josiah Willard Gibbs, an American theoretical physical chemist who conducted most of his experiments in the second half of the nineteenth century

Science History Images Alamy Stock Photo.

From a biochemistry viewpoint, one concept that is fundamental to bio-energetics is free energy and how it is transformed in cells and tissues (see Scientific approach 4.1). To help understand how this term is used in bio-chemistry, let's look again at the molecule that is fundamental to the way that energy is stored and used in cells, ATP (see Figure 3.2). This nucleotide is hugely important because it acts as a biological energy store, being synthesized when some reactions occur and being degraded to help promote others.

Why is ATP so important for metabolism? The answer lies in the large free energy difference that occurs when ATP is hydrolysed—that is, when inorganic phosphate (P_i) is 'broken off' ATP to form adenosine diphosphate (ADP) and an inorganic phosphate ion (P_i), as you saw in Chapter 3. The released P_i is stabilized by forming several chemical species that are just not possible when it is part of the ATP molecule. This seemingly simple biochemical reaction is arguably the key to life and living things.

When ATP is hydrolysed it makes available -30.5 kJ/mole of free energy under what are called 'standard conditions'. The *negative* number means that a significant amount of energy is released—it is made available to the surroundings. In living cells, which don't operate under standard conditions, the free energy change associated with the hydrolysis of ATP is even greater. This is because the concentrations of ATP, ADP, and P_i in living cells are not identical and are lower than those used in standard measurements. In fact, when ATP is hydrolysed in living cells the free energy change ranges from -50 to -65 kJ/mole!

We can measure the change in Gibbs free energy (ΔG or 'dG') for a given reaction (at constant temperature and pressure, of course). It is measured in joules (J) (or kilojoules, kJ). Note that an alternative unit—the calorie—is also used to define energy. Although this is an old, non-SI unit, it is still often used by nutritionists when determining the amount of energy available in food. Look at any food packaging and you will see that the manufacturers provide information about the energy released when the food is metabolized, in either calories or joules—and, often, both. For a reaction to take place spontaneously, the ΔG must be negative; it makes energy available to the surroundings. By contrast, if in theory a reaction has a positive ΔG, then energy must be added to the system to allow it to occur.

A reaction with a positive ΔG cannot proceed spontaneously, but it can still take place, even in biological systems. This is because energy can be provided from other sources (including the Sun, and chemical reactions with a negative ΔG) that are 'coupled' with such reactions. This coupling allows reactions that require energy, such as photosynthesis and DNA synthesis, to proceed without decreasing the total entropy of the universe (which would contravene the second law of thermodynamics). What is amazing about biological reactions is that, even when you take coupled reactions into account, the total entropy (the disorder) in the universe increases. This means that biological systems still follow the second law of thermodynamics (see Figure 4.2).

A brief overview of metabolism

The different sets of chemical reactions of the metabolic pathways in living cells provide the essential resources and energy that the cells need to grow and divide. The amount of energy obtained from these pathways varies depending upon the environment and conditions of the cell. Through some careful analysis, biochemists have identified the different types of reactions that are active in different cells and tissues. What they have discovered is that there is a vast number of reactions, which interact in very complex but highly regulated ways, as shown in Figure 4.3. This diagram highlights how many different reactions can occur in cells, although the exact ones that take place vary for each organism or cell type. You certainly don't need to know them all, but it shows the complexity well.

Each consecutive step in a pathway results in a specific, small chemical change, and the various reactions are exquisitely coordinated to ensure that cells have the biomolecules and energy they need. A small number of metabolic pathways exist in the majority of cells. These have been widely studied by biochemists because they provide information about the general principles of metabolism. Below we briefly describe glycolysis, the series of reactions used by cells when they have good levels of glucose. It is important to remember that sometimes cells need to make their own glucose, which occurs via gluconeogenesis, e.g. when they have low levels of carbohydrates in starvation. These processes are described in Chapter 5.

Although you can't see all of the details within Figure 4.3, it contains two distinct, major types of metabolic pathway. The first type of reaction synthesizes important biomolecules, making polymers from monomers. For example, amino acids are converted to proteins, nucleotides to nucleic acids. These are the anabolic reactions and they usually require an input of energy. The other major type are catabolic reactions. These are the pathways and reactions that break down larger biomolecules into smaller compounds. For example, carbohydrates are broken down into carbon dioxide and water. Generally, catabolic pathways make energy available for cells.

Figure 4.2 Thermodynamics in action on the body of a mole

Anthony Short.

Figure 4.3 This image provides an overview of all metabolic processes that take place in the majority of organisms. Each colour indicates a different metabolic pathway.

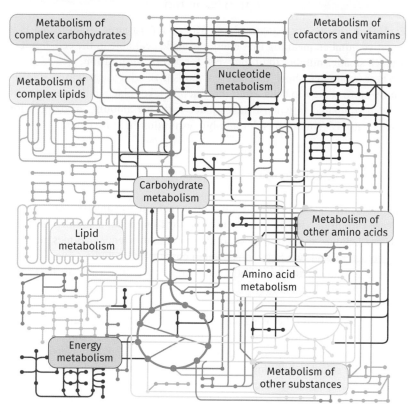

From Berg et al., *Biochemistry* (8th edition, 2015). Macmillan Education.

To find out more about how cells obtain useable energy from catabolic reactions, let's look at cellular respiration. This refers to sets of metabolic reactions that take place in cells to convert energy from nutrients into energy that they can use, producing waste products along the way. Nutrients that are commonly used in respiration include sugars, amino acids, and fatty acids.

Glucose can be used by almost all cells to generate energy, but it can be broken down—catabolized—using different metabolic pathways. Two distinct catabolic reactions involving glucose are aerobic and anaerobic respiration; these release different amounts of energy to cells. Aerobic respiration can only take place when lots of oxygen is available to the cell, with the overall chemical reaction being:

$$C_6H_{12}O_6 + 6O_2 \rightarrow 6CO_2 + 6H_2O \; \Delta G = -2872 \text{ kJ/mole.}$$

As ΔG is a negative number, it means that a large amount of energy is made available to the cell when this reaction occurs. When low amounts of oxygen are available to the cell, however, aerobic respiration cannot take place. Instead, anaerobic respiration occurs. Different anaerobic pathways operate in different cells. For example, during vigorous exercise muscles respire anaerobically to produce lactic acid:

$$C_6H_{12}O_6 \rightarrow 2C_3H_6O_3.$$

Each time this reaction takes place, $\Delta G = -118$ kJ/mole. Again, this is a negative number, meaning that energy is made available to the cell, but it is a much smaller number than that seen for aerobic respiration.

These examples show that the catabolism of glucose releases significantly more energy to cells when the reaction takes place in the presence of oxygen compared to when it takes place with low amounts of oxygen (or none at all). Let's now consider aerobic respiration in more detail.

Aerobic respiration

The basic principle of aerobic respiration is that glucose and oxygen react to release carbon dioxide and water, but it is far from being a simple, one-step process. Cellular respiration incorporates more than twenty enzymes, numerous cofactors, and two ingenious protein complexes, including one with the molecular equivalent of a robotic arm and another like a motor driven by protons that produces molecules of ATP—the overall goal of respiration. This section will give an overview of the process of respiration, including a running commentary on the rising balance of ATP as the process proceeds.

Glucose has to be transformed into a form of energy the cell can readily use. While glucose itself is a good energy *store*, both in food and as glycogen, ATP is much better suited to fulfilling the immediate energy demands inside an active cell. For example, many reactions in metabolic pathways are directly fuelled by the hydrolysis of ATP to form ADP. The spent molecules of ADP are then 'recharged' by phosphorylation as a by-product of the oxidation of glucose.

Five phases make up the overall process of aerobic respiration. Here we will give a very brief description of these phases, with diagrams showing you the processes involved.

Phase 1: glycolysis

Phase 1 is glycolysis, in which six-carbon glucose is split into two molecules of three-carbon pyruvate. This happens in a series of nine steps (see Figure 4.4) in the cytoplasm of the cell. Glucose is first phosphorylated, to stop it escaping from the cell, and then transformed into fructose, ready for the next stages in which it is phosphorylated again, and then split. Both phosphorylation reactions are driven by the hydrolysis of ATP. At this stage, the process of respiration has actually used up two ATP molecules, giving a running balance of −2 ATP.

Many important intermediate compounds are released during the continued oxidation of the three-carbon fragments. In a further five steps, each catalysed by a different enzyme, a pair of pyruvate molecules are produced, and NAD+ is converted into NADH. ADP is phosphorylated to give a total of four ATP molecules, bringing the running balance to +2 ATP. The NADH produced carries electrons to drive vital processes in phase 4 of respoiration (see below).

Figure 4.4 Glycolysis takes place in the cytoplasm of the cell and transforms each molecule of glucose into two molecules of pyruvate, producing NADH and ATP as well. Both of these diagrams represent the same process, although one is much more complex than the other.

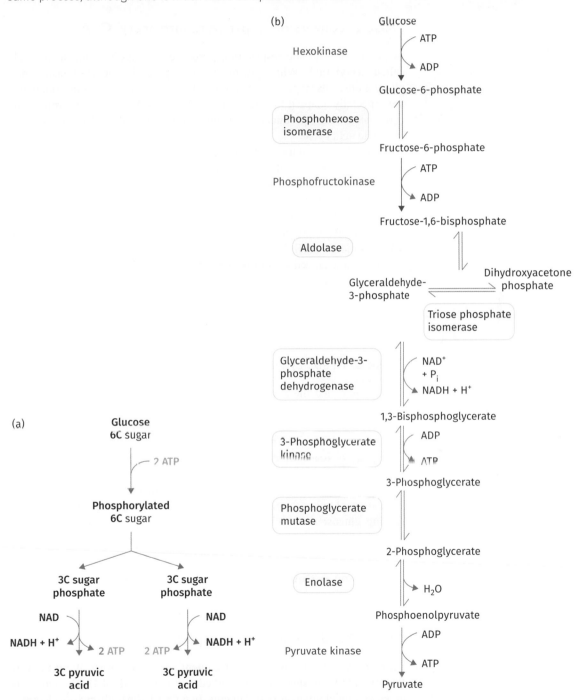

(b): Adapted with permission from Papachristodoulou, D., Snape, A., Elliott, W. H., and Elliott, D. C. (eds), *Biochemistry and Molecular Biology*, Oxford: Oxford University Press, 2018. Copyright © 2018. Reproduced with permission of the Licensor through PLSclear.

The overall chemical reaction for phase 1 is:

$$glucose + 2ADP + 2P_i + 2NAD^+ \rightarrow pyruvate + 2NADH + 2H^+ + 2H_2O + 2ATP.$$

Phase 2: conversion of pyruvate into acetyl CoA

In phase 2 of aerobic respiration, pyruvate is converted into a molecule called acetyl CoA. While glycolysis takes place in the cytoplasm, from phase 2 onwards respiration can only take place in the mitochondria, which are especially adapted for this purpose. Acetyl CoA plays a vital role in the conversion of the products of glycolysis into molecules that can feed into the citric acid cycle.

Although the summary of the reaction looks straightforward, this complicated process demands a sophisticated medley of proteins. The *pyruvate dehydrogenase complex* has a moving arm that physically moves the transforming substrate between the trio of enzymes that compose the complex. This accelerates the overall reaction rate while also reducing the potential for unwanted side reactions. Another by-product, as shown in the equation below, is more NADH for use in phase 4. This process is irreversible and ATP levels are unchanged during phase 2 reactions. This reaction is sometimes referred to as the link reaction.

The overall chemical reaction for phase 2 is:

$$pyruvate + CoA + NAD^+ \rightarrow acetyl\ CoA + CO_2 + NADH + H^+.$$

Phase 3: Krebs cycle (the citric acid cycle)

Phase 3 is called the Krebs cycle (also known as the citric acid cycle). In this stage, pyruvate metabolites are broken down into carbon dioxide, and high-energy electrons are harvested. At the beginning of the cycle, acetyl CoA reacts with an organic compound called oxaloacetate to make citrate. While the oxaloacetate is regenerated during the cycle, the incoming acetyl CoA is transformed into two molecules of carbon dioxide. Important by-products from the cycle include one molecule of ATP (or GTP, a related good store of energy), bringing the running ATP balance up to +4 for each starting glucose; the cycle also yields three more molecules of NADH and another of $FADH_2$, which is also a carrier of electrons (see Figure 4.5).

The overall reaction of the citric acid cycle is:

$$Acetyl\ CoA + 3NAD^+ + FAD + ADP + P_i + 2H_2O \rightarrow 2CO_2 + 3NADH + FADH_2 + ATP + 2H^+ + CoA.$$

Phases 4 and 5: oxidative phosphorylation

For many years, there was much argument among scientists about the way that oxidative phosphorylation worked (see Case study 4.1). The currently accepted model involves two elements. In phase 4, the high-energy electrons are transferred to molecular oxygen. This leads not only to the production of water, but also to the creation of a proton gradient across a key membrane, the **inner mitochondrial membrane** (Figure 4.6(a)). Finally, in phase 5,

Figure 4.5 The citric acid cycle. Acetyl-CoA combines with oxaloacetate at the start of the citric acid cycle. The subsequent steps use electron carriers to harvest high-energy electrons, while also releasing ATP. Oxaloacetate is regenerated but the acetyl-CoA is metabolized to carbon dioxide.

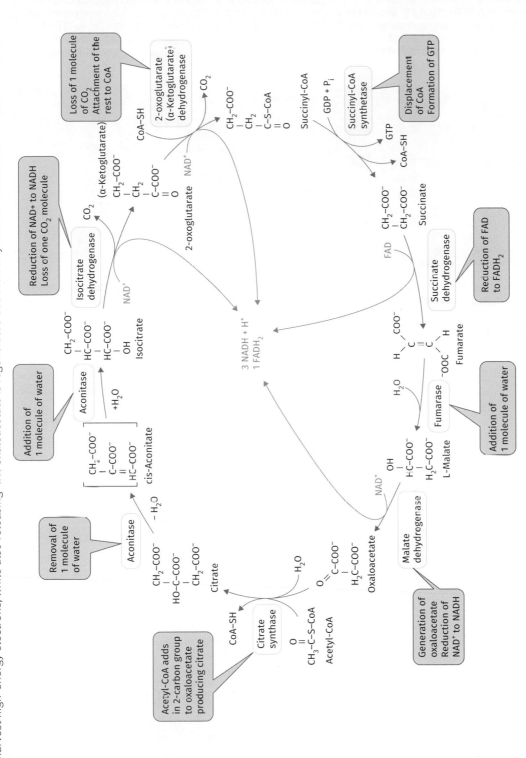

Adapted with permission from Papachristodoulou, D, Snape, A., Elliott, W. H., and Elliott, D. C. (eds), *Biochemistry and Molecular Biology*, Oxford: Oxford University Press, 2018. Copyright © 2018. Reproduced with permission of the Licensor through PLSclear.

Figure 4.6 The electron transport chain occurs in mitochondrial membranes. (a) Mitochondria contain an inner membrane that separates the matrix from the intermembrane space. The inner membrane is extensively folded and compartmentalized, as can be seen in this electron micrograph. (b) The inner membrane contains a chain of proteins that transfer high-energy electrons through to oxygen molecules. This chain of proteins is called the electron transport chain. Further details are given in Figure 4.8.

(a)   (b)

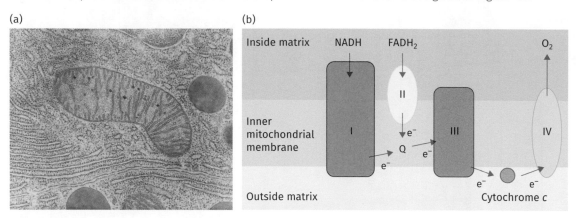

(a): Science History Images/Alamy Stock Photo; (b): adapted with permission from Papachristodoulou, D., Snape, A., Elliott, W. H., and Elliott, D. C. (eds), *Biochemistry and Molecular Biology*, Oxford: Oxford University Press, 2018. Copyright © 2018. Reproduced with permission of the Licensor through PLSclear.

a rotating protein complex traps energy from the proton gradient to create the vast majority of the ATP molecules liberated by aerobic respiration.

During phase 4, the electron carriers from earlier phases, notably the NADH, transfer their high-energy electrons to three of four membrane-bound proteins, through which the electrons take a complicated route via a series of iron and copper ions (Figure 4.6(b)). These ions can exist in various oxidation states, which makes them well suited to this role of electron transport, since they can accept and donate electrons readily via reversible reactions, like so:

$$\text{Iron:} \quad Fe^{3+} + e^- \leftrightarrow Fe^{2+}$$

$$\text{Copper:} \quad Cu^{2+} + e^- \leftrightarrow Cu^+.$$

Eventually, these electrons will reduce molecular oxygen into water molecules. In the meantime, however, their passage through the membrane proteins is coupled to the deposit of protons (hydrogen cations, H^+) on the outside of the inner mitochondrial membrane, while removing them from the inside. This leads to an all-important difference in proton concentration (known as a proton gradient) across the membrane—the basis of the chemi-osmotic theory developed by Mitchell (see Case study 4.1).

The overall equation for phase 4 is:

$$2NADH + O_2 + 2H^+ \rightarrow 2H_2O + 2NAD^+.$$

Phase 5 is the stage that delivers the ATP needed by all the cells of the body. The proton gradient across the inner mitochondrial membrane is important: it results in an energy difference (properly called an energy potential) across the membrane, which is used to drive the generation of huge quantities of ATP. Specifically, the higher concentration of protons on the outside of the membrane creates both a concentration gradient and a charge gradient. In tandem these gradients are described as the **proton-motive force**. Scientists were mystified for many years as to how this could drive the phosphorylation of ADP to produce ATP until biochemists Paul Boyer and John Walker worked out the details (see Case study 4.1).

A protein complex called **ATP synthase** (see Figure 4.7) phosphorylates multiple molecules of ADP. The membrane-bound complex is composed of a barrel that rotates relative to a stationary base. The barrel comprises six polypeptides: three alpha and three beta subunits, arranged in alternating fashion. The rotation of the barrel relative to the stationary base causes the beta subunits to cycle between open and closed conformations. Initially, in the open conformation the active site binds a molecule each of ADP and inorganic phosphate. As the barrel rotates, the active site enters the closed conformation, during which ADP is phosphorylated. Entering the open conformation once more, the resulting molecule of ATP is ejected, allowing for the whole process to start again.

A cascade of protons across the membrane spins the motor to drive ATP synthesis. ATP synthase harnesses the proton-motive force via a special ion channel, composed of two half-channels that do not meet in the middle (see Figure 4.8). Instead, protons enter a half-channel, then ride the rotating

Figure 4.7 ATP synthase: a complex enzyme which is key to the production of ATP in cells

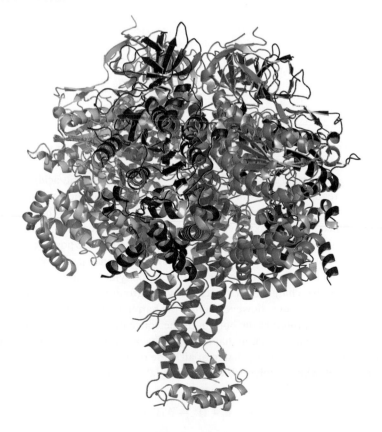

Figure 4.8 A proton-motive force drives the synthesis of ATP by ATP synthase. High-energy electrons are transferred via a chain of proteins containing metallic ion cofactors to oxygen molecules. This has the effect of increasing the relative concentration of protons in the intermembrane space of the mitochondria, which are used by ATP synthase through the following steps. (1) Protons move into a half channel in the ATP synthase complex, where they attach to amino acid residues in the rotating barrel of the synthase complex. (2) The alternating attachment and release of protons causes the barrel to rotate. (3) Rotation of the barrel relative to the base of the synthase complex causes a cyclical conformation change, which catalyses the phosphorylation of ADP to produce ATP.

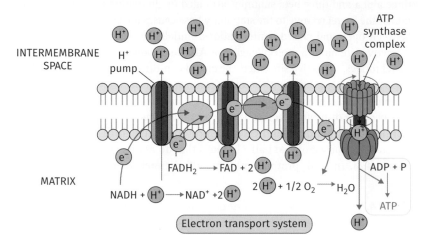

Adapted from Kaiser, G. (2019). 17.5: Phosphorylation Mechanisms for Generating ATP. [online] Biology LibreTexts. Available at: https://bio.libretexts.org/Bookshelves/Microbiology/Book%3A_Microbiology_(Kaiser)/Unit_7%3A_Microbial_Genetics_and_Microbial_Metabolism/17%3A_Bacterial_Growth_and_Energy_Production/17.5%3A_Phosphorylation_Mechanisms_for_Generating_ATP. Distributed under the terms of the Creative Commons Attribution 4.0 International (CC BY 4.0). https://creativecommons.org/licenses/by/4.0.

barrel like a carousel to reach the other half-channel, through which they complete their journey.

Amino acids such as glutamic acid (or sometimes aspartic acid) can be found at regular intervals around the circumference of the rotating barrel. Each of these residues contains a carboxylic acid group that can be deprotonated to its carboxylate form. In their deprotonated form, these negatively charged carboxylate groups will be attracted to the positive potential of the protons outside the inner membrane. When a proton enters the half-channel it forms a bond to one of these carboxylate groups and neutralizes its negative charge. Now the carboxylate group is no longer attracted to the external protons; however, across the barrel another such residue will be dumping its proton at the bottom of the concentration gradient, when it will regain a negative charge that attracts it back towards the outside of the membrane. The alternating charge of all these residues drives the revolution of the barrel, which catalyses the phosphorylation of ADP to liberate ATP.

The overall reaction for phase 5 is:

$$ADP + P_i + H^+ \rightarrow ATP + H_2O.$$

Case study 4.1
Making waves: how an understanding of electron transport chains caused controversy

There is now general agreement about the biochemical mechanisms that underpin energy transduction in different cells, but this has not always been the case. The development of some of the hypotheses described in this chapter was troublesome. Ultimately, though, for the biochemists involved, there were huge rewards as their proposals became accepted.

Peter Mitchell developed the chemiosmotic theory for respiration that is outlined in this chapter, but his rise to stardom revealed an ugly side of science. The research biochemist won the Nobel Prize in Chemistry for the hypothesis that ultimately explained how respiration produces ATP. In 1961, he shared his ideas in the journal *Nature*, when he coined the term *proton-motive force*. The originality of his views, especially as an outsider to the field of bioenergetics, was too much for his peers. Like many scientists before him, he experienced extreme hostility to his hypothesis, at one point so harsh that he had to take two years off because of stomach ulcers.

To buoy his spirits, he kept a chart on which he monitored the exact dates each rival capitulated to his views. A critic who initially wrote off Mitchell's ideas as 'ridiculous and incomprehensible' went on to produce one of the most compelling pieces of evidence in its support. André Jagendorf and his colleague Ernest Uribe suspended chloroplast membranes in acidic solution. When alkali was added, they observed the production of ATP. When Mitchell was awarded the Nobel Prize for Chemistry in 1978, he quoted the physicist Max Planck, who said: 'A new scientific idea does not triumph by convincing its opponents, but rather because its opponents eventually die.' In his acceptance speech, Mitchell called the defiance of this principle a 'singularly happy achievement'. His ideas extend well beyond human biochemistry, explaining phenomena in plants and even bacteria, which exploit the proton-motive force for a variety of processes, including the rotation of their propelling flagella.

As protons move back into the mitochondrial matrix by chemiosmosis, the molecular machine known as ATP synthase generates ATP. The way that this enzyme works also generated much discussion and debate. Eventually, it was landmark studies of the ATP synthase that were recognized in the 1997 Nobel Prize for Chemistry that was awarded to Paul D. Boyer and John Walker. Conclusions from biochemical studies undertaken by Boyer were originally controversial because they suggested that energy input to the enzyme is not used to form ATP directly. Instead, the energy promotes the binding of inorganic phosphate and the release of bound ATP at three identical catalytic sites and their sequential binding changes are driven by the rotation of a smaller internal subunit. This novel mechanism was eventually supported by structural studies by John Walker.

These final two phases of respiration deliver up to twenty-six more molecules of ATP for each original glucose—the majority of the ATP produced during aerobic respiration. Overall, each molecule of glucose metabolized by aerobic respiration can produce at least thirty molecules of ATP.

We have seen how the respiration of glucose is a fantastically complex process that achieves an important transformation. A by-product of the oxidation of glucose to form carbon dioxide and water is the liberation of ATP, which fuels numerous processes at the cellular level.

Photosynthesis

Track back far enough in time and we will find that almost all the energy contained within terrestrial life originates from our Sun, either directly or indirectly. In this chapter, we have until now reviewed different aspects of metabolism where energy is converted between different organic forms. Now it's time to turn our attention to the biological systems that emerged millions of years ago to exploit the energy that leaves the sun as sunlight, and is captured by living cells using a variety of pathways termed photosynthesis.

We all know that sunlight splits into a spectrum of different wavelengths of light, each at different energies (Figure 4.9(a)). Some cells extract energy at particular wavelengths by using specific molecules that literally capture light. In recent years, as humans have sought alternatives to fossil fuel, they have endeavoured to mimic this process by using solar panels to provide for some of our energy needs (Figure 4.9(b)). These technologies have relatively low efficiency for capturing solar energy in a form that is useful. This leaves open a potential role for biochemists, who may be able to engineer biological approaches to improve the technologies that are currently available to us (find out more in Chapter 7).

Most plants, algae, and cyanobacteria perform photosynthesis; such organisms are called photoautotrophs—they make their own food in the form of organic molecules using light. Photosynthesis is largely responsible for producing and maintaining the oxygen content of the Earth's atmosphere, and it supplies all the organic compounds and most of the energy required for life on Earth. So how do plants and microbes capture the energy that is used to fuel their biological activities?

Before we consider the detail, let's set out an overview of photosynthesis. Photosynthesis occurs in two stages. In the first stage, light-dependent reactions (or light reactions) capture the energy from sunlight and use it to make the energy-storage molecules ATP and NADPH. In the second stage, light-independent reactions (those that don't need sunlight to proceed) use these products to capture and reduce carbon dioxide to carbohydrates. It takes six water molecules and six carbon dioxide molecules to form $C_6H_{12}O_6$ (simple glucose) and six oxygen molecules. The process is summarized in Figure 4.10, and the general equation for the miraculous reaction of photosynthesis is:

$$CO_2 + H_2O + photons \rightarrow [CH_2O] \text{ (carbohydrate)} + O_2.$$

In other words, carbon dioxide and water combine (with the input of energy) to generate carbohydrate (sugar) and oxygen.

Figure 4.9 Light contains a spectrum of wavelengths and energies. (a) This schematic diagram demonstrates that sunlight is made of electromagnetic radiation of different wavelengths and energies. Biological photosynthetic reactions absorb light in the range 400–700 nm. (b) Humans now routinely capture the energy from sunlight through use of solar panels.

(a)

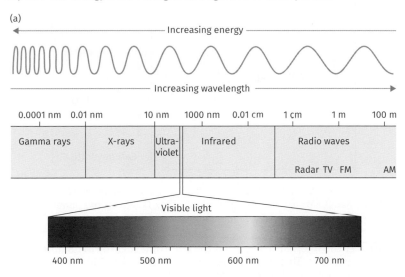

(b)

(b): Love Silhouette/Shutterstock.com.

Let's now consider the light-dependent and light-independent reactions in turn.

Although photosynthesis is performed differently by different species, the process always begins when proteins called reaction centres absorb light. In plants, the reaction centres contain green chlorophyll pigments held inside chloroplasts (Figure 4.11), which are most abundant in leaf cells.

Figure 4.10 The main stages of photosynthesis.

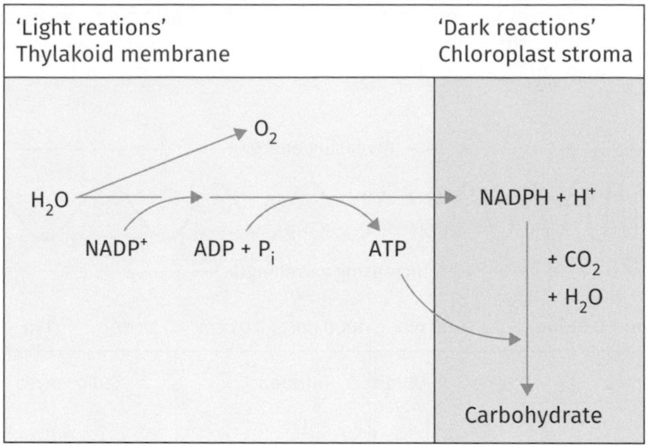

'Light reations' Thylakoid membrane	'Dark reactions' Chloroplast stroma

O_2

H_2O

$NADP^+$ $ADP + P_i$ ATP

$NADPH + H^+$

$+ CO_2$
$+ H_2O$

Carbohydrate

Adapted with permission from Papachristodoulou, D., Snape, A., Elliott, W. H., and Elliott, D. C. (eds), *Biochemistry and Molecular Biology*, Oxford: Oxford University Press, 2018. Copyright © 2018. Reproduced with permission of the Licensor through PLSclear.

Figure 4.11 The internal structure of the chloroplast is key to its functions in photosynthesis. This structure is shown magnified 13,000 times in this electron micrograph at 10cm on the long side.

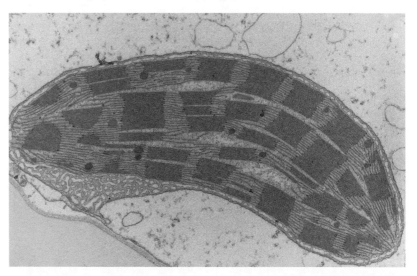

Science History Images/Alamy Stock Photo.

In bacteria, such as the beautiful blue-green cyanobacteria, these reaction centres are embedded in the plasma membrane.

In these light-dependent reactions, some energy is used to remove electrons from suitable cellular substances, such as water. The reaction produces oxygen and hydrogen cations (as protons). The hydrogen freed when water molecules split is used to create two further compounds that act as short-term energy stores, which can be drawn on to fuel other reactions. One of these compounds is our old friend ATP, but there is another biological molecule that can also play this role: it is the reduced form of nicotinamide adenine dinucleotide phosphate, or NADPH for short.

In plants, algae, and cyanobacteria, long-term energy storage takes the form of the production of sugars. These sugars are produced by a subsequent sequence of light-independent reactions called the Calvin cycle; some bacteria use different mechanisms, such as the reverse Krebs cycle, to achieve the same end. In the Calvin cycle, atmospheric carbon dioxide is incorporated into already existing organic carbon compounds, such as ribulose-1,5-bisphosphate (also known as RuBP). The resulting compounds are then reduced, using the ATP and NADPH produced by the light-dependent reactions, to form further carbohydrates, such as glucose. One of the enzymes involved in these first steps of photosynthesis is ribulose-1,5-bisphosphate carboxylase/oxygenase, commonly known as Rubisco. This enzyme catalyses the carboxylation of RuBP and is probably the most abundant enzyme on Earth.

The first photosynthetic organisms probably evolved early in the history of life on Earth and most likely had to use reducing agents such as hydrogen or hydrogen sulfide, rather than water, as sources of electrons. Cyanobacteria (Figure 4.12) appeared later; and the vast excess of oxygen they produced contributed directly to the oxygenation of planet Earth. As humans, we owe everything to the cyanobacteria that allowed the evolution of complex lifeforms that rely on a steady supply of oxygen. This oxygenic photosynthesis is by far the most common type of photosynthesis and the overall process of oxygenic photosynthesis in plants, algae, and cyanobacteria is quite similar. As well as oxygenic photosynthesis, there are many varieties of anoxygenic photosynthesis, which are used by certain types of bacteria. In this case, the bacteria use light to make sugars from carbon dioxide but they do not release oxygen.

Many photosynthetic organisms are photoautotrophs: they can synthesize organic compounds directly from carbon dioxide and water using energy

Figure 4.12 A light micrograph of *Tolypothrix*, a type of Cyanobacteria

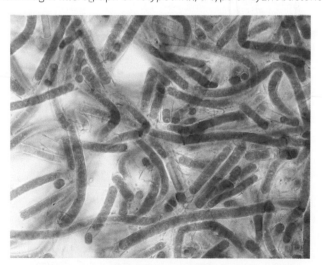

Peter Siver, Visuals Unlimited/Science Photo Library.

from light. However, not all organisms use carbon dioxide as their carbon source to carry out photosynthesis; photoheterotrophs use more complex organic compounds, instead of carbon dioxide, as a source of carbon.

Carbon dioxide is converted into sugars using the energy captured from sunlight in a process called carbon fixation. Carbon fixation is an endothermic redox reaction—in other words, it requires energy. In general, photosynthesis is the opposite of cellular respiration: cellular respiration is the oxidation of carbohydrate or other nutrients to carbon dioxide, while photosynthesis is a process of the reduction of carbon dioxide to carbohydrate. This tells us clearly that photosynthesis and cellular respiration are distinct processes, with different energy requirements; they take place using different sequences of chemical reactions; and they occur in different cellular compartments.

As we see in the general equation for photosynthesis a little earlier, water is essential for photosynthesis. So it is perhaps not surprising that the first microbes to undertake photosynthesis were probably found in marine environments. Water is both a substrate in reactions that are light-dependent and a product of reactions that are light-independent. Most organisms that use oxygenic photosynthesis exploit visible light for the light-dependent reactions. Leaves contain chlorophyll pigments that absorb blue and red light but reflect green wavelengths, which is why they appear green to us (Figure 4.13). Some organisms exploit other pigments for photosynthesis that can use shortwave infrared or far-red radiation.

Figure 4.13 Absorption spectra of chlorophyll a (blue) and chlorophyll b (red) in a solvent

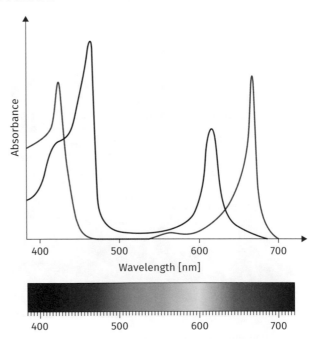

Variations on a theme

The type of photosynthesis described until now is the form used by the majority of land plants—it is also known as C3 photosynthesis because it generates phosphoglycerate, a three-carbon compound. However, this is not the only version of photosynthesis. There are at least two other forms in plants alone, illustrated in Figure 4.14. Bacteria and archaea (ancient, single-celled organisms that are similar to but fundamentally different from

Figure 4.14 Summary of the different photosynthetic pathways. C3 plants fix CO_2 directly, using the enzyme ribulose bisphosphate carboxylase oxygenase (Rubisco). C4 plants use PEP carboxylase to react CO_2 with phosphoenolpyruvate (PEP) to eventually form malate, which is transferred to the bundle sheath, where it is broken down to release CO_2 that is used by Rubisco. CAM plants use PEP carboxylase to fix CO_2 at night and then break down the malate produced during the day to provide CO_2 for Rubisco.

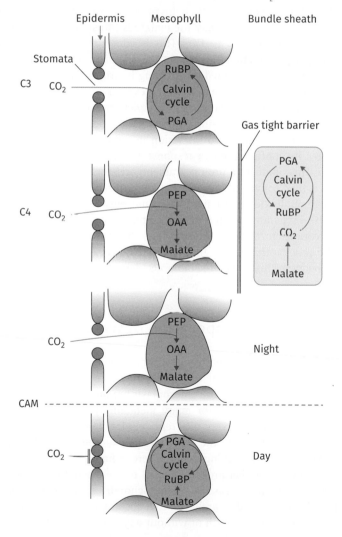

bacteria) have other, very different biochemical pathways that make photo-synthesis possible. Here is a brief insight into some of these variations—and why they are important.

C4 photosynthesis takes its name because of the four-carbon molecule, malate, that is the first product of this type of carbon fixation. It is a more elaborate form of C3 photosynthesis and is likely to have evolved more recently. C4 photosynthesis ensures that Rubisco works in an environment where there is a lot of carbon dioxide and little oxygen. Additional biochemical reactions are involved, but they require more energy in the form of ATP to fix carbon in drought, high temperatures, and limitations of nitrogen or CO_2. Since the more common C3 pathway does not require this extra energy, it is more efficient in other conditions.

Another variant of photosynthesis is crassulacean acid metabolism, or CAM for short (Figure 4.14). Plants with this metabolism keep the stomata in the leaves shut during the day to reduce evaporation, but open at night to collect CO_2. The enzyme phosphoenolpyruvate (PEP) carboxylase fixes CO_2 at night, and then during the day the malate is transported to chloroplasts, where it is converted back to CO_2. This is concentrated around Rubisco, increasing photosynthetic efficiency in the use of water. This makes CAM plants particularly well adapted for growing in arid conditions.

Some prokaryotic organisms employ even more radical variants of photosynthesis. Some archaea use a simpler method of photosynthesis that involves bacteriorhodopsin, a three-chain protein that is similar to a protein pigment complex used for vision in animals. The bacteriorhodopsin changes its configuration in response to sunlight, and acts as a proton pump. This produces a proton gradient, and the energy stored in these gradients is then converted into ATP by ATP synthase. The process does not involve carbon dioxide fixation and does not release oxygen, and seems to have evolved separately from the more common types of photosynthesis.

These various forms of photosynthesis demonstrate that it is a vital metabolic process that drives life: it captures the light used to drive the anabolic reactions that create the biomolecules required by all living things.

Endosymbiosis: cooperation between organisms

Studies of respiration and photosynthesis have helped us understand some of the most fundamental concepts about how life on Earth evolved. As we have seen, mitochondria and chloroplasts are membrane-bound cellular organelles that are essential for many metabolic processes in eukaryotic cells. Although they have very different cellular functions, they have one thing in common: they are the result of endosymbiosis. Both organelles are the result of an intriguing event where a host cell—an original pro-karyote—absorbed another prokaryotic cell to the mutual benefit of both—an endosymbiosis! In the case of the mitochondrion, this one-off event is thought to have involved an archaeon that absorbed a facultative anaerobic bacterium (which can live with or without oxygen). Similar types of events led to the original chloroplasts, but in this case it was likely to have been a cyanobacterium that was absorbed.

These endosymbiotic events happened at least 500 million years ago, when oxygen levels were low and metabolic pathways in the progenitor cells would have been relatively inefficient for generating energy. The early eukaryotic-like cells that emerged had the advantage of having a more diverse metabolism.

Several clues led to the hypothesis of mitochondria and chloroplasts having resulted from endosymbiosis. Both organelles contain circular genomes—like other bacteria that still exist on the Earth. They are also able to self-replicate. Lynn Margulis first postulated this hypothesis back in the 1960s, but the idea wasn't well received by the scientific establishment. It took fifteen rejections from different scientific journals before finally her theory was published. These days, the theory is widely accepted because experiments have shown that the chloroplast genome has a striking resemblance to genomes of cyanobacteria.

Eukaryotes are genetic chimeras: they have inherited genes from their archaeon hosts, but they also possess genes they have gained from their endosymbiont. Over time, many genes from the chloroplast and the mitochondrion transferred to the host genome and are still used to produce the proteins required for these organelles to function effectively in the cells. Research published in 2018 has provided further support for this theory, with scientists having engineered a laboratory model for this type of evolutionary milestone. The recent experiments created a yeast dependent for energy on bacteria living inside it as a beneficial parasite or 'endosymbiont'.

In this chapter, we have shown that biochemical studies of cellular metabolism have helped us to understand the diversity of life on Earth, and the likely origins of its more complex forms.

Chapter summary

- The sum of all processes within each cell that cover the synthesis, degradation, and inter-conversion of different molecules is referred to as its metabolism. There are many different sets of pathways, but a relatively small number are fundamental to all cells.

- A full appreciation of cellular metabolism requires a clear understanding of how energy is transferred between different molecules and made available to drive reactions.

- Two distinct, major types of metabolic pathways occur in all cells. Anabolic reactions synthesize important biomolecules and usually require an input of energy. Catabolic reactions break down complex biomolecules into smaller compounds and make energy available for cells.

- One theoretical concept that is useful is that of free energy, which identifies the energy available from a reaction to do work.

- ATP is fundamental for all cells through its role in acting as a store for energy, helping to link favourable reactions with those that would not normally take place in cells.
- Aerobic respiration is a set of biochemical reactions in which glucose and oxygen react to release carbon dioxide and water. It starts with glycolysis, and progresses to the citric acid cycle. It involves more than twenty enzymes and numerous cofactors, eventually enabling the electron transport pathway to generate large amounts of ATP in the presence of oxygen.
- Photosynthetic pathways exploit the energy available in sunlight, using specific proteins to extract energy from particular wavelengths. Various forms of photosynthesis exist, demonstrating that these vital metabolic processes drive the anabolic reactions that create the biomolecules required by all living things.
- Endosymbiosis describes the process where one prokaryotic cell absorbed another to the mutual benefit of both. These events led to the formation of the earliest types of mitochondria and chloroplasts, eventually leading to the wide diversity of complex life that now exists on Earth.

 Further reading

http://www.mrc-mbu.cam.ac.uk/projects/2248/molecular-animations-atp-synthase: a series of animations based on high-resolution molecular structures that describe how electrons pass along the respiratory chain from one protein to the next and generate a proton gradient across the mitochondrial membrane.

Margulis, L. (1999) *Symbiotic Planet: A New Look at Evolution*. Phoenix, London.

https://pdb101.rcsb.org/motm/27: this gives a neat description of bacteriorhodopsin and the process of photosynthesis in archaea.

 Discussion questions

4.1 Make a careful comparison of the processes of aerobic respiration and photosynthesis. Discuss why there is more potential for using photosynthesis to solve the world's energy problems than there is to use aerobic respiration.

4.2 If plants can use energy from the Sun to make ATP, what are the advantages of using it up to make glucose just so they can make ATP again via respiration?

4.3 Dinitrophenol has been used as a slimming aid, although it is now illegal for it to be sold and used for human consumption. Investigate how this compound impacts on aerobic respiration and discuss both why it might appear to be a useful aid in weight control and why it is so dangerous.

MAINTAINING A METABOLIC BALANCE

5

A healthy lifestyle requires careful and efficient production of all the molecules we have met in this book so far. One job of metabolism is to ensure that the right balance of all of these molecules is maintained, a process known as **homeostasis**. A common example includes the systems that maintain a constant concentration of glucose in the bloodstream. This chapter will investigate those and other processes in more detail, along with the biomolecules that help to achieve it. Although similar systems are found in many living creatures, we will focus on the systems found in the human body.

It is vital that our bodies maintain a healthy balance of all the biomolecules met in the earlier chapters. We need carbohydrates, such as those which act as cell markers and as fuel for cellular respiration; lipids, like those which form the cell membrane; proteins, including those needed to build muscle tissues; and nucleic acids, including the building blocks of DNA. We get what we need from our food, which provides us not only with the building blocks to produce important biomolecules, but also with vital substances that assist with construction and metabolism, including micronutrients. These are substances of which we only need up to 100 milligrams per day, including trace elements and vitamins. This is why a balanced diet is so important: if we don't get the nutrients we need, our body doesn't function properly (Figure 5.1).

Figure 5.1 From the moment of your birth, your body has to take in everything it needs by eating or breathing

PosiNote/Shutterstock.com.

What makes a balanced diet?

A balanced diet should contain suitable amounts of a variety of nutrients to sustain a healthy and fully functioning body. Starches, like rice, bread, and pasta, provide glucose. Fish, meat, and legumes are rich in proteins, which the body disassembles into amino acids to act as building blocks for our own proteins. Of the twenty standard amino acids our bodies need, almost half are described as essential because we lack the ability to make them ourselves. A notable example is methionine, deficiency of which plays a key role in a disease called kwashiorkor. This disease is common among children affected by a lack of protein in their diet and causes a distended (bloated) belly. Similarly, our bodies can produce lipids but are unable to produce certain essential fatty acids, such as omega-3, which is found in fish and many vegetables. As we will see, the body is able to produce many nutrients from alternative sources: glucose can be produced from proteins as well as fats.

Vitamins

The main role of vitamins is to make proteins work. Just as a superhero often needs a sidekick, proteins often need an activating particle called a cofactor. These particles, which can be molecules or monoatomic ions, bind to proteins to enable the special functions we expect from them.

One important cofactor gives us the power of sight. Retinal is derived from vitamin A and it binds to a family of proteins called rhodopsins, which enable us to see (see Figure 5.2). A double bond in retinal absorbs

Figure 5.2 Eyes are amazing structures, but without vitamin A they don't work very well. Retinal is a co-factor for the rhodopsin proteins that enable us to see.

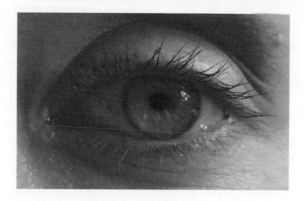

Anthony Short.

light in the visible region of the electromagnetic spectrum. When this happens the π-bond breaks, allowing rotation around the bond, which causes a conformational change in the rhodopsin. (See Chapter 2 to remind you of the role of rhodopsin in the eye.) No pure sequence of amino acids has evolved which can carry out this function without the assistance of retinal. Instead, we have to make sure that we eat foods like carrots, yellow peppers, and eggs, which provide us with vitamin A in the form of retinol. In the body retinol is transformed into retinal, for use with rhodopsins and other metabolic processes. Eating insufficient vitamin A can lead to problems with your eyesight, including night blindness and corneal damage, as well as damage to the respiratory and gastrointestinal tracts.

Another vital vitamin is B3. Without it we would be unable to access the energy we need to live. Also known as niacin, this vitamin plays an important role in glycolysis, a series of reactions which enable us to extract energy from glucose, as you saw in Chapter 4. Niacin is found in yeast, meat, cereals, milk, green leafy vegetables, and even tea and coffee. In the body niacin is transformed into nicotinamide adenine dinucleotide, usually referred to as NAD^+. This coenzyme (a cofactor which activates an enzyme) is one of many examples of an **activated carrier**, biomolecules that carry energy or molecular groups to aid metabolism. In this case, NAD^+ acts as an electron bridge, accepting electrons from glucose and passing them onto molecular oxygen. Deficiency of vitamin B3 leads to a condition called *pellagra*, which causes skin problems, diarrhoea, and depression.

Another example of an activated carrier is vitamin B5. Also known as pantothenic acid, the vitamin comes from foods including liver, kidney, yeast, egg yolk, broccoli, and sweet potatoes. In the body it is converted into an activated carrier called coenzyme A (CoA). Rather than transferring electrons, like NAD⁺, CoA transfers acyl groups in a variety of metabolic processes, including the synthesis and oxidation of fatty acids, which can be used to make cell membranes or provide energy. Deficiency in vitamin B5 can lead to hypertension, characterized by high blood pressure, which increases the risk of serious conditions like stroke and heart attacks.

Trace elements

Trace elements are an important kind of micronutrient. They are absorbed in the form of polyatomic ions and include metals, like copper and calcium, and non-metals, such as iodine and selenium (see Scientific approach 5.1). Like vitamins, they often act as cofactors to proteins.

Iodine is a good example of a trace element vital for healthy metabolism. It is required to make two hormones produced in the thyroid gland, located in the neck, called thyroxine and triiodothyronine. People often talk about having a 'fast' or 'slow' metabolism. These thyroid hormones play a major role in controlling that rate, changing how quickly carbohydrates are absorbed and fatty acids are released. Unsurprisingly, a common symptom if your thyroid gland is underactive is to gain weight, whilst if it overproduces hormones you will lose weight fast!

Iodine is found in white fish, shrimp, tuna, eggs, and dairy products. In many countries, including the UK, iodine is commonly added to table salt, which has proved to be an effective way to prevent iodine deficiency. In spite of this, according to World Health Organization data, iodine is one of the top three micronutrients from which people suffer deficiency around the world. Deficiency leads to hypothyroidism, with symptoms including extreme fatigue, goitre (Figure 5.3), mental slowing, depression, and weight gain.

Figure 5.3 If the thyroid gland does not get enough iodine, it cannot make enough thyroid hormone. This is detected and the gland grows and works harder to try and make more. This results in a characteristic swelling in the neck, known as a goitre.

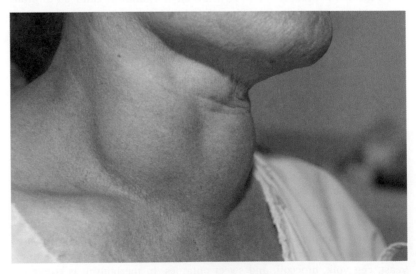

Karan Bunjean/Shutterstock.com.

Scientific approach 5.1
Selenium—a miracle cure?

Selenium is a vital partner to iodine. When the body uses iodine to make thyroxine and triiodothyronine, the chemical reactions produce free radicals as a harmful by-product. Left unchecked, these highly reactive species can damage DNA and cause mutations in the genome. Luckily, the body has evolved protective mechanisms that mop up these dangerous radicals; selenium is a co-factor of glutathione peroxidase, an enzyme which transforms the radical hydrogen peroxide into water. Selenium is found in many foods, including seafood, various meats, eggs, and nuts—Brazil nuts in particular. Selenium deficiency has been linked by several studies to various types of cancer, including prostate, colorectal, and lung cancer.

Two landmark selenium trials recorded very different results.

In 1996, the world celebrated an apparent cure-all for cancers of every shade that is, except skin cancer. The moderately sized trial, with 1312 participants, investigated the impact of selenium on skin cancer. Members of the treatment group took 200mg a day of a special kind of yeast rich in selenium (Figure A). Although the micronutrient had no effect on skin cancer, participants had far fewer cases of prostate, lung, and colorectal cancers, with 63 per cent, 58 per cent, and 45 per cent fewer cases affecting the treatment group compared to the placebo group, respectively. The results were so promising that doctors ended the trial early so that all participants could start to benefit from selenium supplements.

Unfortunately, a randomized follow-up trial involving 32,400 men in 2009 posed difficult questions for this apparent panacea. Originally designed to

Figure A High-selenium yeast was the selenium supplement used in the 1996 trial

Nedim Bajramovic/Shutterstock.com.

last twelve years, this trial was also cut short after seven years after no benefits of taking selenium were detected. Its results are summarized in Figure B. In fact, selenium was linked with a higher risk of developing type 2 diabetes, although the data were not significant, meaning the higher incidence of diabetes could simply have been by chance and nothing to do with the selenium.

These clashing results are hard to fully explain. The 1996 trial selected participants with a selenium deficiency, whereas the 2009 trial selected healthy individuals. The first trial probably demonstrates the dangers of being deficient in selenium, rather than its miraculous healing properties. However, the second trial used a different form of selenium supplements (L-selenomethionine, the amino acid methionine with selenium in place of sulfur; Figure C), which could also have affected the results.

? Pause for thought

How do you think researchers decide at which stage it has become unethical to withhold treatment from the placebo group?

Figure B The 1996 selenium supplement study recorded a higher incidence of both tested kinds of skin cancer in the treatment group, but much lower rates of all-cause mortality (death for any reason), total cancer mortality (death caused by any variety of cancer), and total cancer incidence (developing any form of cancer, whether or not fatal), and specifically much lower incidence of colorectal, lung, and prostate cancers.

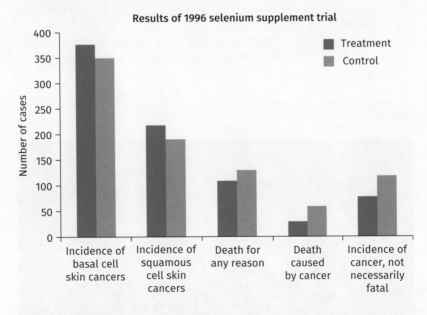

Figure C The 2009 trial used a different selenium supplement, L-selenomethionine, complicating the issue of comparing data from the two trials

An example from history involves Switzerland, where a serious epidemic of hypothyroidism and goitre gripped the country for many years. As Switzerland is landlocked, its citizens had much less access to sea fish than their neighbours, depriving them of a valuable source of iodine. Consequently, iodine deficiency was virtually unavoidable at the turn of the twentieth century, with nearly 100 per cent of school children suffering with goitre, a cricket ball-sized swelling of the glands in the neck. When the need for iodine was identified, the element was added to table salt and the incidence of goitre and related problems, such as severe learning difficulties and stunted growth, disappeared rapidly.

Copper is an essential mineral—but deadly in excess. Vomiting, diarrhoea, and stomach pain can result if humans drink water with more than 6 parts per million of copper, which is about 0.11 mg/kg. Copper is found in legumes, mushrooms, chocolate, liver, nuts, and seeds. It is a cofactor for cytochrome C oxidase, an enzyme which alters the charge gradient across cell membranes to enable the synthesis of the energy vector ATP in chemiosmosis (see Chapter 4). The enzyme is essential for the body to absorb iron from food and to incorporate it into haemoglobin, enabling the passage of oxygen around the body. Copper deficiency can lead to osteoporosis, which causes weak bones, poor metabolism of cholesterol and glucose, and problems with blood pressure, heart function, and immunity.

Trace elements, like vitamins, are needed in tiny amounts but carry out vital functions in the body. A healthy balanced diet should reflect this.

The glucose/glycogen balancing act

The process of cellular respiration is described in Chapter 4. You need glucose all the time to fuel this vital process—but you only take in food to supply that glucose a limited number of times in a day. So a key aspect of your metabolism is managing the surges of glucose flooding into the blood after a meal, while making sure that a stockpile of glucose is readily available for when the blood glucose levels fall. One element of this metabolic balancing act is the conversion of glucose to glycogen (see Chapter 1 for details of these compounds).

Glycogen synthesis

Glucose is converted into glycogen in the liver by four different enzymes. The first enzyme, *UDP-glucose pyrophosphorylase*, transforms glucose into an activated form called UDP-glucose. The second enzyme, *glycogenin*, acts as a seed for the glycogen, by linking together four molecules of UDP-glucose, one of which stays docked to the *glycogenin*. At this stage the third enzyme, *glycogen synthase,* begins adding additional UDP-glucose monomers to the growing chain. Once biomolecular monomers have been polymerized, they can be referred to as residues. When the chain is at least eleven residues long, a fourth enzyme chops off strands roughly seven residues in length and reattaches them elsewhere, to promote branching (see Figure 5.4).

Figure 5.4 Glycogen synthesis

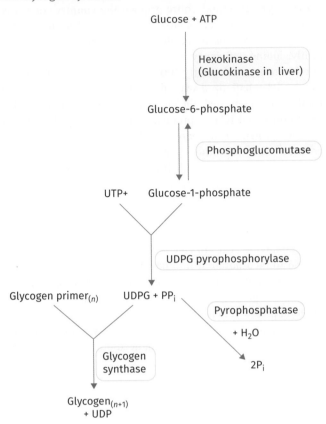

As so often happens in biochemistry, even this pathway is a simplification. The production of UDP-glucose itself is not a single-stage reaction—it also involves a series of steps. First of all, glucose is phosphorylated in the 6-position by adenosine triphosphate (ATP). Next, the phosphate group is transferred from the 6- to the 1-position. Finally, the glucose 1-phosphate reacts with a nucleotide called uridine triphosphate (UTP), producing the UDP-glucose and a pyrophosphate group (which is basically two phosphate groups bonded together) (Figure 5.5).

The role of insulin in glycogen synthesis

Glycogen synthesis is overseen by the hormone insulin. The main job of the pancreas is to monitor levels of glucose. When the concentration of glucose is high the pancreas will release insulin, which signals the liver and muscle cells to start storing glucose as glycogen. This is achieved via two different signal cascades.

The first effect of insulin is to usher more glucose into cells. The liver acts as a general depot for glucose, but glycogen synthesis also takes place in muscle cells, so that glucose is immediately available during exercise. Insulin promotes the uptake of glucose into these cells by signalling for the

Figure 5.5 The enzyme UDP-glucose pyrophosphorylase transforms glucose-1-phosphate into UDP-glucose

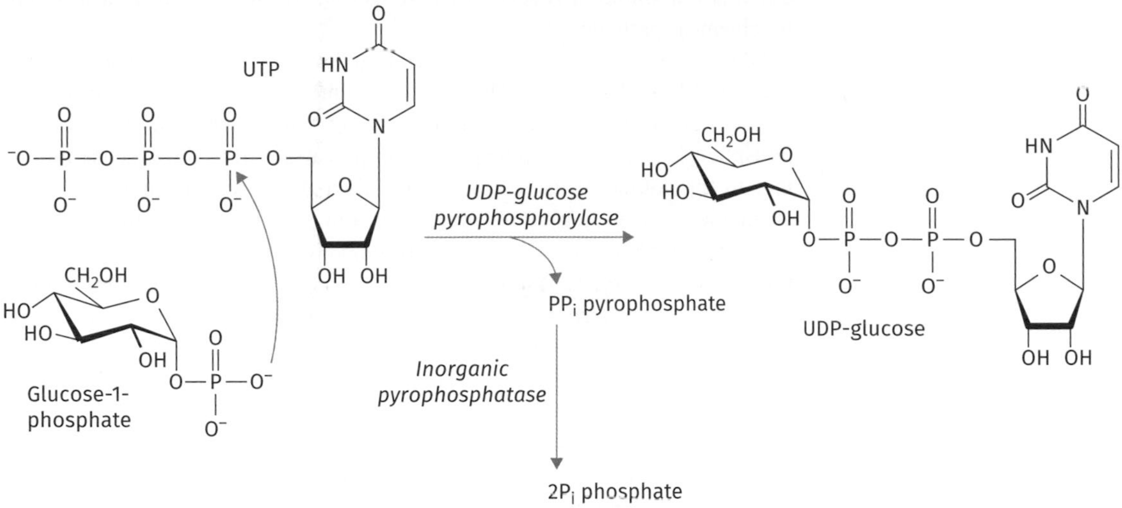

release of GLUT transporters. These proteins, which convey molecules of glucose across the cell membrane, are kept on standby in large sacs called storage vesicles. Insulin binds to receptors, which send the signal for these GLUT molecules to be released into the cell membrane, increasing the rate at which glucose is moved into the cells.

The second effect of insulin is to activate glycogen synthase. Recall that this enzyme does the heavy lifting with the polymerization of glucose to form glycogen. When blood sugar levels are low, the enzyme is put on standby in a deactivated state. A phosphate group is attached to the glycogen synthase, which slightly changes its shape and hence prevents it from doing its job. Confusingly, this deactivation is carried out by an enzyme with the similar name of glycogen synthase kinase. Both enzymes are deactivated by the attachment of a phosphate group. Insulin triggers a signal cascade that phosphorylates glycogen synthase kinase, which conversely dephosphorylates glycogen synthase, liberating it to polymerize glucose into glycogen.

Glycogen synthesis is essential for healthy body function. When the process fails, the body is forced to expel excess glucose in urine, meaning that this invaluable fuel is wasted. As we shall see, this failure is usually the result of a breakdown in the important signalling carried out by insulin.

Releasing glucose from storage

The opposite of glycogen synthesis is glycogen breakdown. When blood sugar levels are low, the body releases glucose from storage in the process called glycogen breakdown.

Exercise and fasting are the key triggers for glycogen breakdown. If a person goes some time without eating, blood sugar levels will fall. The pancreas identifies this shortage and signals the liver to release glucose from glycogen stores. Signals are also released shortly before and during exercise, as this increases the need for glucose in muscles. Glycogen break-down is not simply a reversal of glycogen synthesis—it follows a different biochemical pathway.

Like glycogen synthesis, the breaking down of glycogen also requires four different enzymes. The first enzyme, glycogen phosphorylase, removes glucose monomers from the glycogen polymer chain. It is called a phosphorylase because a phosphate group is attached to the glucose mono-mer in the course of its removal from glycogen. Glycogen phosphorylase is unable to remove glucose monomers once it gets within four residues of a branch. Consequently, two more enzymes are required to constantly remodel the glycogen. When a branch has been reduced to four monomers in length, a *transferase* enzyme cuts off a strand of three monomers and reattaches it to the main chain. But this leaves a stump, which has to be removed by *α-1,6-glucosidase*. This releases a single molecule of glucose, which is then phosphorylated by the fourth enzyme *hexokinase* (Figure 5.6).

The amount of glycogen in our muscles and liver can affect our athletic performance—but does it also affect how we learn? We've all been there: we should be studying, but instead we're thinking, 'I'll just give my room one last tidy up.' No student's bedroom is cleaner and tidier than during revision season. But why? Athletes often 'hit the wall', a feeling of fatigue that maps closely to the depletion of glycogen stores in muscles. But the same is not true of mental exertion. Although the brain uses one-fifth of all the energy we take in, it doesn't use much more whether you're watching television or struggling to solve a tricky maths problem. So while scientists are still unsure why mental effort is such a spectre, it's not because it drains fuel supplies. This also means you can't lose weight by worrying.

Glycogen breakdown is regulated by a small army of biomolecules. The process is initiated by two different hormones, glucagon and adrenaline.

Hormonal control of glycogen breakdown

Glycogen breakdown is set in motion by the release of glucagon and adren-aline (or epinephrine, as it is usually known in the US). Glucagon is most active in the liver, while adrenaline mostly triggers glycogen breakdown in muscle cells. The job of the liver is to maintain a steady concentration of blood sugar levels around the whole body. Consequently, low blood sugar levels trigger the release of glucagon from the pancreas. Meanwhile, adren-aline is released during exercise, and in response to danger, from the adrenal medulla gland at the top of the kidney. This is part of the *fight-or-flight* response, and it is often called an adrenaline rush (although the response is characterized by many more signalling molecules than just adrenaline). When both hormones trigger the same cell, the overall rate of glycogen breakdown increases.

Adrenaline and glucagon activate signal cascades in target cells (see Figure 5.7). The hormones bind a membrane-bound receptor called a G-protein, which works with a medley of additional proteins to transform large numbers of molecules of the energy vector ATP into cyclic adenosine monophosphate (cAMP). This is called a signal cascade because a single molecule of glucagon or adrenaline leads to the production of many mol-ecules of cAMP.

Figure 5.6 (a) Glucose residues are sequentially removed from the shrinking glycogen chain. Different enzymes are required to lop branches and remove stumps. This constant remodelling enables the complete transformation of glycogen to separate molecules of glucose. (b) Glycogen branches are linked via different positions on the ring. Branches link via a 1,6 glycosidic linkage, whereas straight chain residues are linked via 1,4 glycosidic linkages.

(a)

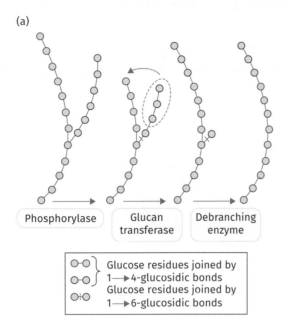

| Phosphorylase | Glucan transferase | Debranching enzyme |

Glucose residues joined by 1⟶4-glucosidic bonds
Glucose residues joined by 1⟶6-glucosidic bonds

(b) Two different enzymes are needed to cleave the respective 1–4 and 1–6 glycosi linkages in glycogen

CH$_2$OH   CH$_2$OH

Amylo-1,6-glucosidase, cleaves the alpha 1–6 linkage

α[1–6] linkage

CH$_2$OH   CH$_2$OH   CH$_2$ CH$_2$OH

α[1–4] linkages

Glycogen phosphorylase cleaves the alpha (1–4) linkages

(a): From Rodwell, V W. et al. (Eds.). *Harper's Illustrated Biochemistry* 13/e: McGraw Hill Education. (b): Adapted with permission from Papachristodoulou, D., Snape, A., Elliott, W. H., and Elliott, D. C. (eds), *Biochemistry and Molecular Biology*, Oxford: Oxford University Press, 2018. Copyright © 2018. Reproduced with permission of the Licensor through PLSclear.

Figure 5.7 (1) Glucagon hormone activates membrane-coupled G-protein. (2) Activated G-protein stimulates the release of cyclic adenosine monophosphate (cAMP). (3) The signal cascade continues with the activation by phosphorylation of protein kinase A (PKA). (4) PKA activates phosphorylase kinase (PPK) by phosphorylation. (5) PPK activates glycogen phosphorylase (PYG) by phosphorylation, thereby promoting glycolysis. (6) PKA also stimulates the gluconeogenesis pathway while down-regulating the glycogen synthesis pathway.

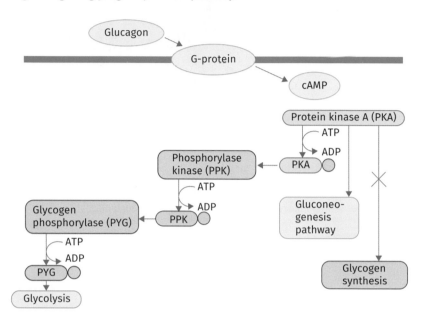

The release of cAMP molecules initiates a sequence of events via three enzymes: cAMP activates *protein kinase A*, which activates *phosphorylase kinase*, which activates *glycogen phosphorylase*. As with glycogen synthesis, there is much opportunity for confusion because the enzymes pass phosphorylase groups around rapidly. Both phosphorylase kinase and glycogen phosphorylase are activated by phosphorylation, but also activate their substrate by phosphorylating it.

The first stage leads to the activation of phosphorylase kinase. The cAMP molecules released from the signal cascade bind to and activate protein kinase A, which then partially activates phosphorylase kinase by attaching a phosphoryl group to it.

Phosphorylase kinase is a highly sophisticated enzyme. Formed from a total of twelve subunits, the protein is composed of two identical lobes, each of which contains three regulatory sites and one active site, bridged by the remaining four subunits. On each lobe, two of the subunits are activated by phosphorylation and one is activated by calcium ions. Muscles contract when signals from the brain flood cells with calcium ions. As well as activating the proteins that cause the muscle cells to contract, this calcium also activates the phosphorylase kinase. Although liver cells do not contract, they also contain stores of calcium ions, which are released in response to adrenaline signalling. With six regulatory subunits, phosphorylase kinase is more like a volume dial than a light switch. The more of the regulatory units that are activated, the more effectively the enzyme works.

The final step enables glycogen breakdown. Phosphorylase kinase activates glycogen phosphorylase. Both enzymes have analogous relationships with the phosphoryl group. Both are activated by the attachment of a phosphoryl group, while both names contain the word 'phosphorylase', denoting that each also activates its substrate by phosphorylating it. Once activated, phosphorylase kinase A attaches a phosphoryl group to glycogen phosphorylase, which then begins removing glucose residues from glycogen, releasing molecules of the sugar in its phosphorylated form.

Glycogen phosphorylase makes sure that glycogen synthesis does not happen at the same time as glycogen breakdown. While the enzyme is activated by phosphorylation, it is also able to bind molecules of glucose, which deactivates it. If there is already a lot of glucose in the cell, it would be a waste of energy to break down glycogen, so it is a very useful adaptation that glucose molecules can directly deactivate glycogen phosphorylase.

Energy wastage is also kept to a minimum by hormonal activity. Glycogen breakdown only occurs when adrenaline and/or glucagon is binding a cell's receptor. As soon as the hormones are released, the signal cascade halts. Meanwhile, all cells contain **phosphodiesterases**, which remove cAMP from the cell, bringing a stop to the activities of all the enzymes downstream from protein kinase A. This ensures that glycogen breakdown only occurs when the energy is needed for exercise, or to top up blood sugar levels. Similar systems ensure that glycogen synthesis only occurs when blood sugar levels are high, so the body wastes no energy storing glucose for which there is an immediate need.

By balancing glycogen synthesis with glycogen breakdown, the body maintains a steady blood sugar level, while also making fuel rapidly available during exercise.

Inborn errors of metabolism

Problems with genes can lead to problems with metabolism. Sometimes when we inherit mutated genes from our parents, we cannot make working versions of vital proteins. The resulting problems with metabolism might cause anything from mild discomfort to death, if left undiagnosed. Our growing understanding of these congenital disorders has improved the lives of many people. A modified diet can be enough to avert what would have been a death sentence 150 years ago.

Archibald Edward Garrod was the first scientist to realize that metabolic conditions could be inherited from parents. In 1908, he coined the term 'inborn error of metabolism' while detailing insights into a medical condition called **alkaptonuria**. Patients passed black urine and developed a blue tint in the hollows of their ears and brown marks on the membrane covering their eyes. Garrod associated these symptoms with two amino acids, tyrosine and phenylalanine. Both contain phenyl groups, which he linked to the difficulty these patients had with metabolizing the nutrients.

What made Garrod's insight even more remarkable was his realization that alkaptonuria was homozygous recessive. Another such trait is the inheritance of a curved thumb, which only occurs when children inherit the genes responsible from both parents. If a child inherits the alleles for both curved and straight thumbs, their thumb will be straight, which is the dominant characteristic. Garrod's insights were remarkable because he knew nothing of Gregor Mendel, the pioneering monk who discovered dominant and recessive traits by cross-breeding peas. In spite of this, Garrod realized that alkaptonuria would only result when children inherited a recessive gene from both parents. His landmark research led the way to a rich body of science, in which congenital disorders of metabolism are increasingly treatable.

- Alkaptonuria is caused by problems with an enzyme called homogentisate 1,2-dioxygenase. This is one of several enzymes that break down both tyrosine and phenylalanine into the end product succinylacetone. The metabolic pathway of both amino acids is almost identical because phenylalanine can be transformed into tyrosine by a single enzyme. So any problems metabolizing tyrosine will also affect phenylalanine. Two more enzymes convert tyrosine into homogentisic acid, which is where problems arise. This metabolite is normally broken down by homogentisate 1,2-dioxygenase, but mutations in the gene render the corresponding enzyme ineffective, leading to a build-up of homogentisic acid. Excess levels of this metabolite are flushed from the body in urine, which it turns black. The acid also causes ochronosis, which darkens the tissue of cartilage and bone, while also damaging joints and cardiac valves.

- Phenylketonuria is a condition related to alkaptonuria. Affected individuals cannot metabolize phenylalanine but they can break down tyrosine. As we saw earlier, the first step in the metabolism of phenylalanine is its conversion into tyrosine. In phenylketonuria, this is not possible because a genetic mutation impairs the relevant enzyme, *phenylalanine hydroxylase*. As a result, phenylalanine builds up in fluids all over the body with potentially life-threatening consequences. Three-quarters of affected people who are not diagnosed die by the time they are thirty.

- Brain damage is the main symptom of phenylketonuria. Although the exact mechanism is unclear, scientists know that the same protein that transports tyrosine is also able to deliver phenylalanine and tryptophan across cell membranes. Too much phenylalanine could cause traffic jams, depriving neurones of tyrosine and tryptophan, which are precursors for the neurotransmitters dopamine and serotonin, respectively. Dwindling stocks of neurotransmitters may also explain other problems, including reduced myelin sheaths around the neurones. Myelin helps neurones signal more effectively, so lack of myelin leads to reduced nervous activity.

- Unlike alkaptonuria, phenylketonuria has a potent remedy. Placing patients on a low phenylalanine diet allows healthy development of brain tissue. Proteins which are already low in phenylalanine, like

Figure 5.8 Phenylketonuria is one metabolic disorder with a relatively simple solution, through a simple blood test carried out in the first few days of life and a low-protein diet

Lewis Houghton/Science Photo Library.

the casein in milk, are treated to remove the amino acid altogether. In one study, patients who started the diet within a few weeks of birth recorded an average IQ of 93 compared to control group members, who scored an average of 53 after starting the diet up to a year after birth. In the UK, all parents are offered a heel prick blood test of their new born babies to identify affected infants and treat them immediately—giving them the opportunity of a healthy, relatively unaffected life (see Figure 5.8).

Starvation vs obesity

The amount of food we take in varies greatly—by choice or necessity. Many cultures practise fasting, when people limit how much food they eat—e.g. Muslims fast during the month of Ramadan. In some countries, conditions are such that people simply never have enough to eat. During such fasting and even during starvation, the body has biochemical systems which maintain healthy levels of glucose. Conversely, the West has long been in the grip of an obesity epidemic, which is spreading to countries around the world as diets change. Scientists are satisfied that the reasons for this epidemic are not mere greed. Many inborn errors of metabolism have been identified which promote obesity, often by hijacking our appetites so that we never feel full. A different set of reactions manage the metabolism when we take in more calories than we need.

Gluconeogenesis

When the body does not get enough glucose from the diet, there are bio-chemical pathways to produce it by other means. Our brains alone require 120g of glucose a day and if the sugar is missing from our diet, our bodies generate it from other sources in a process called gluconeogenesis, which literally means *making new glucose*. There are three different feedstocks for this process, including amino acids. You may have been warned that going on a calorie-restricted diet can lead to a loss in muscle mass. This is the source of these amino acids, which are produced from the hydrolysis of proteins in our muscles. Another intriguing source is lactic acid, produced as a result of anaerobic respiration, which is the cause of the burning sensation when bouts of exertion leave our muscles with insufficient oxygen. This lactic acid can be converted back into glucose in the liver. Finally, the best known alternative source is fat, or more specifically its glycerol backbone. (Triacylglycerols are explained in more detail in Chapter 2.) Gluconeogenesis takes place mainly in the liver, but also in the kidneys and the brain.

Gluconeogenesis achieves the reverse of glycolysis via alternative pathways. In glycolysis, glucose is transformed into pyruvate, whereas gluconeogenesis produces glucose from pyruvate. However, gluconeogenesis is not a simple reversal of glycolysis for two reasons: first, the different feedstocks enter the gluconeogenesis pathway at different stages, and second, four different enzymes are used.

We can picture gluconeogenesis as a ladder with ten rungs. On the ground is the starting material for the whole process, pyruvate, which can be transformed into glucose via ten enzyme-catalysed steps. Some of the feedstocks can skip one or more of these rungs, by direct transformation into one of the intermediates on the gluconeogenesis pathway.

There are three types of feedstock in the gluconeogenesis pathway—lactate, amino acids, and glycerol. Lactate is the ion remaining when the lactic acid produced during exercise releases its acidic hydrogen ions. The liver converts lactate into pyruvate, which has to climb all ten rungs of the gluconeogenesis ladder. Some amino acids, such as alanine and glycine, also start at this stage. The first step of gluconeogenesis is the conversion of pyruvate into oxaloacetate, the chemical that combines with acetyl CoA in the first step of the citric acid cycle. Other amino acids, such as aspartate and asparagine, are converted directly into oxaloacetate, so they can skip straight to the first rung of the ladder. The final feedstock, glycerol, starts directly on the sixth rung in the form of glyceraldehyde 3-phosphate, one of the key metabolites we met in glycolysis in 'The glucose/glycogen balancing act' earlier.

Alternative enzymes offer a way out of thermodynamic dead ends. All catalysts, including enzymes, reduce the activation energy in both directions of a chemical reaction. Most of the enzymes used in glycolysis can also be used in gluconeogenesis, because the shortage of glucose and glycolytic intermediaries drives equilibrium in the direction of gluconeogenesis. But three steps in glycolysis are so exergonic they can only be reversed by using different enzymes. Basically, gluconeogenesis is almost a straight reversal of glycolysis, except an additional step is added and four different enzymes are employed to overcome thermodynamic resistance (see Figure 5.9).

Figure 5.9 Gluconeogenesis employs four enzymes that are different to those in glycolysis (highlighted here in orange), and one extra step (highlighted in blue)

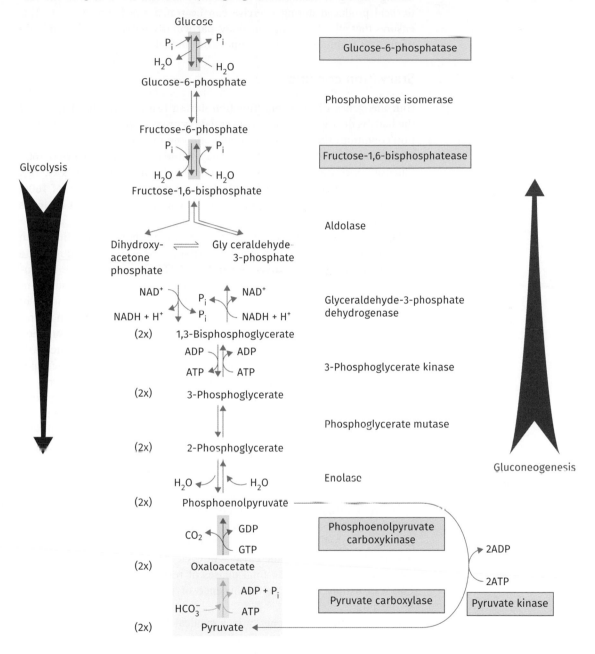

Gluconeogenesis provides the body with an alternative source of glucose during fasting and starvation. In this way protein, fat, and even the lactic acid produced during exercise can be transformed into glucose. This ensures that all cells around the body, particularly red blood cells and brain tissue, can maintain adequate supplies of the fuel.

Starvation conditions

Gluconeogenesis is a normal function during fasting but what happens if the body is deprived of food for several days or even weeks? In such starvation conditions the body shifts to maintain energy supplies.

The brain uses 120g of glucose per day, while the whole body needs 160g glucose. Since the brain is said to use one-fifth of all the body's energy, these figures might seem confusing because 120g is three quarters of 160g, not one-fifth! This is because many of our energy needs are met by lipids (see Chapter 2).

Lipids are a highly effective way to store energy but they are limited in the ways they can make this available. Triacyl glycerides break down into fatty acids and glycerol. Glycerol can be transformed into glucose by gluconeogenesis, but fatty acids cannot. Meanwhile, in starvation conditions fatty acids are metabolized into acetyl CoA, which is also the product of phase 2 of respiration. As we saw in Chapter 4, phase 2 of respiration is irreversible, explaining the inability of fatty acids to restock glucose supplies. Nevertheless, their acetyl CoA metabolite acts as feedstock for the citric acid cycle, driving all the remaining phases of respiration. The distinction is important because red blood cells depend for their only source of energy on glucose. Since red blood cells lack mitochondria, they liberate energy from glucose anaerobically, using a similar pathway to the one used in muscles when they lack sufficient oxygen during exercise. This is an important feature of red blood cells, because otherwise they would consume the oxygen they are adapted to deliver. In normal conditions, the same is true for brain cells, but as we shall see, in starvation conditions the brain can function on a different kind of energy vector called ketone bodies.

Starvation triggers a competition for oxaloacetic acid. Unlike acetyl CoA, oxaloacetic acid can be transformed into glucose via gluconeogenesis. But the acetyl CoA liberated via the metabolism of fatty acids also requires oxaloacetic acid, so that the citric acid cycle can capture the high-energy electrons necessary for the final phases of respiration. Fortunately, acetyl CoA can supply energy in the absence of oxaloacetic acid by forming ketone bodies.

Ketone bodies supply energy in a cyclical fashion. First, acetyl CoA is transformed into three ketone bodies: acetoacetate, D-3-hydroxybutyrate, and acetone (see Figure 5.10). The liver, where this process takes place, is unable to metabolize ketone bodies for energy, so these highly soluble energy vectors are quickly transported around the body. Next, energy is released by the metabolism of D-3-hydroxybutyrate, which yields high-energy electrons in the form of NADH, as well as increasing the proton gradient across the inner mitochondrial membrane. In other words, ketone bodies can skip the citric acid cycle to liberate energy via oxidative

Figure 5.10 Acetyl CoA is metabolized to form three ketone bodies: acetoacetate, acetone, and D-3-hydroxybutyrate (shown here in purple)

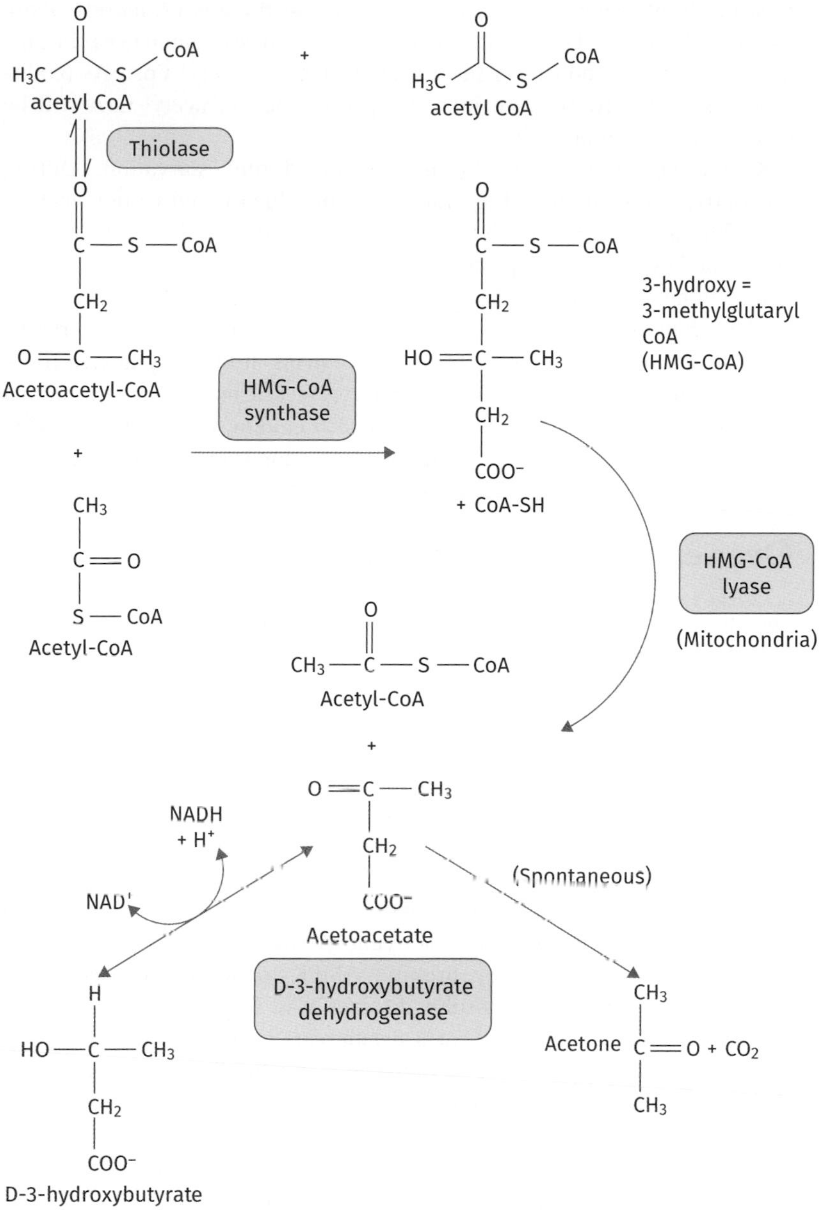

Adapted with permission from Papachristodoulou, D., Snape, A., Elliott, W. H., and Elliott, D. C. (eds), *Biochemistry and Molecular Biology*, Oxford: Oxford University Press, 2018. Copyright © 2018. Reproduced with permission of the Licensor through PLSclear.

phosphorylation. Curiously, the ultimate metabolite is acetyl CoA, meaning the feedstock for the process is regenerated.

If no food becomes available, starvation proceeds via the following steps. First of all, glycogen supplies are exhausted so there is no more glucose available from storage. At this stage gluconeogenesis commences, in tandem with the metabolism of fatty acids to liberate acetyl CoA. As oxaloacetate supplies dwindle, an increasing proportion of acetyl CoA will be converted into ketone bodies.

Ketone bodies maintain brain function during starvation. During non-starvation conditions the brain relies on glucose and cannot metabolize fatty acids for energy, but as starvation continues, the brain shifts to ketone bodies for its energy needs.

Protein metabolism is the final, dangerous stage of starvation. Once all fat has been converted into ketone bodies, the only other option is to convert protein into glucose via gluconeogenesis. All proteins in the body have a function, so cannibalizing them for energy depletes muscle mass, which the animal relies on to find the food it needs, while also eroding the function of vital organs like the brain and heart. Continued starvation leads quickly to death.

Over-eating, obesity, and diabetes

The opposite of starvation is excess intake of dietary calories. During times of plenty the metabolism shifts to store excess energy in the form of lipids, including triacylglycerides.

Many checks and balances ensure that lipid synthesis only takes place when the body has surplus energy. Regulation hinges around the enzyme acetyl CoA carboxylase 1, which helps to transform the acetyl group from acetyl CoA into fatty acids. High concentrations in the cytoplasm of ATP and acetyl CoA—both indicators that energy is abundant—indirectly activate acetyl CoA carboxylase 1 via the enzyme AMP-activated kinase (AMPK).

Hormones and diet also help regulate lipid synthesis. Glucagon and adrenaline, the hormones which activate glycogen breakdown, also suppress fatty acid synthesis by priming AMPK to deactivate acetyl CoA carboxylase 1. Conversely, insulin disables AMPK to promote fatty acid synthesis. Finally, changing what a person eats influences the genes they express. Quitting starvation for a high-carb, low-fat diet promotes expression of the genes linked with lipid synthesis.

Continually eating too many calories upsets these balances with unhealthy consequences. In fact, obesity has been strongly linked with type II diabetes. Both types of diabetes mellitus, named from the Latin for honey, *mel*, are caused by a failure of insulin signalling. Type 1 diabetes results from an inability of the body to produce sufficient amounts of working insulin, which can arise from a variety of genetic mutations. Type 2 diabetes results when the body stops responding to insulin, a process known as insulin resistance. This is when the body sends insulin signals which are ignored. No one is yet sure why this happens, but there are several research directions.

All forms of acute diabetes are characterized by a spike in blood glucose. One of the functions of insulin is to signal muscle, liver, and adipose cells (adapted for the storage of fat) to uptake glucose. Failure of this signalling causes all cells to reject the sugar and raid lipids for energy. Meanwhile, adipose cells send the products of lipid breakdown to the liver in the form of glycerol, which is then transformed into glucose by gluconeogenesis, and fatty acids. Consequently, the body produces more and more glucose, even though the sugar is already abundant in the bloodstream. In the acute phase of the condition the excess blood glucose signals the kidneys to flush the sugar out in urine, leading to unusual levels of hunger and thirst and numerous trips to the bathroom.

Failure to treat diabetes poses lethal dangers via ketone bodies. Lipid breakdown in adipose cells raises concentrations of both glycerol and fatty acids in the liver. These fatty acids are metabolized to produce acetyl CoA. Unfortunately, in the absence of oxaloacetate, which is depleted by the ongoing gluconeogenesis, this excess of acetyl CoA is transformed into ketone bodies. Since there is no shortage of glucose in the body, they are not metabolized for energy. Unfortunately, ketone bodies are acidic so their continued accumulation reduces the body's pH, impairing function, which leads to coma and then death. Recall that proteins are supremely sensitive to pH, meaning that a loss of pH balance has the ability to cause a conformation change in proteins with an associated loss of function. This is particularly problematic in the brain and untreated diabetics often seem drunk, because of impaired brain function and the ketone body acetone on their breath, which is mistaken for the smell of closely related alcohol (ethanol).

Insulin resistance and its role in type 2 diabetes has proved extremely difficult to characterize. One of the problems is that the body often has an alternative pathway to reach the same goal. Scientists often test the function of proteins by removing the gene entirely from the genome of animals under investigation. One experiment knocked out the insulin receptor gene in mice, yet found them able to uptake glucose in the targeted muscle cells during exercise. Earlier, in 'The glucose/glycogen balancing act', we saw how insulin signals muscles to uptake glucose by mobilizing molecules of GLUT, the transporter which ferries the sugar across cell membranes. Studies have found this same protein is mobilized when muscles contract, via AMPK, the kinase which helps to regulate lipid synthesis. These competing pathways make it very hard to understand exactly how insulin resistance works.

Many correlates of insulin resistance have been uncovered. Researchers have identified a potential role for a different kind of lipid—diacylglyceride (DAG), which have two fatty acids per glycerol backbone instead of three. When concentrations of DAG are high in muscle tissue, certain enzymes are activated which counteract insulin signals. For example, where insulin signalling might phosphorylate one of the downstream enzymes in the signal cascade, another biochemical pathway might dephosphorylate the same enzyme. So while molecules of insulin signal muscle cells to absorb more glucose, other agents block entry.

Obesity is another noted correlate. People can present with insulin resistance who are not obese, but it is much more common in people who are. Does this mean that obesity causes insulin resistance or could insulin resistance cause obesity? A more specific correlate is fatty liver, where the liver develops too much fatty tissue. Dietary choices could also be a factor.

Fructose used to be recommended to people on diets because it is sweeter, mass for mass, than glucose. But as we saw earlier, fructose is an intermediary in the glycolysis pathway, which gets hijacked when the sugar becomes too concentrated in the liver. Accordingly, dogs fed a high-fructose diet in one study went on to develop insulin resistance.

Much work remains to characterize the potential link between obesity and diabetes. While a causal relationship is yet to be established, it seems like good advice to eat a balanced diet with a suitable number of calories. Luckily there are many effective treatments for both types of diabetes, as well as several exciting research leads to combat the disease and obesity itself (see Case study 5.1).

Case study 5.1
Genes and appetite

Problems with appetite could be a major factor in obesity. Carriers of a mutated form of the receptor gene MC4R were found to eat more dietary fat in a study involving curry. The receptor is activated by a hormone called melanocortin in the hypothalamus, a region of the brain linked with appetite.

Carriers of the mutated gene preferred higher-fat dishes even when they tasted no better than lower-fat equivalents. The 2016 study, led by Cambridge Professor Sadaf Farooqi, hinged on three chicken korma dishes, which looked, smelled, and tasted identical, but contained different amounts of fat. Participants were given samples of the curries, which contained 20 per cent, 40 per cent, and 60 per cent fat respectively, and asked to rate their palatability. Next they were invited to eat as much of whichever curries they pleased until they felt comfortably full. Although there was no significant difference in the flavour ratings, the MC4R-deficient participants were found to eat much more of the high-fat korma than two control groups, composed of lean and obese individuals, respectively.

While the study was very small, involving just fifty-four participants, the results suggest that genetic mutations could influence a person's ability to eat a suitable amount of food for their lifestyle. Another interesting finding was that the MC4R-deficient participants demonstrated a much lower preference for sugar in a similar experiment involving the dessert Eton Mess (see Figure A).

❓ Pause for thought

How does a small study size limit the quality of scientific data?

Figure A Taste preference vs intake: participants rated the flavour of the three different kormas, then helped themselves to as much of each dish as they liked. After the meal, the curries were rated again. Graphs A and B show that taste scores were comparable for the low fat (20 per cent), medium fat (40 per cent), and high fat (60 per cent) curries, while graphs C and D demonstrate a clear preference for the highest fat curry among the participants deficient in the MC4R gene.

Chapter summary

- A balanced diet should include correct proportions of macronutrients, vitamins, and trace elements. Starches provide fuel in the form of glucose, which powers cellular processes around the body. Dietary proteins and fats form building blocks for our own proteins and lipids, as well as an alternative fuel supply. Vitamins and trace elements are often cofactors which enable proteins to do their jobs.

- Levels of glucose are homeostatically regulated by the hormones insulin, glucagon, and adrenaline. When glucose is abundant it is stored in the form of glycogen. When glucose is in demand, such as during exercise, glycogen is broken down.

- Inborn errors of metabolism result from genetic mutations. Enzymes which are either ineffective or in too short supply allow substrates to build up to potentially harmful levels.

- Under starvation or fasting conditions the body can generate its own glucose via the gluconeogenesis pathway, which transforms glycerol, certain amino acids, and oxaloacetate into glucose. Lipids also provide a supply of acetyl CoA, some of which is transformed into the alternative energy vector: ketone bodies.

- In response to an excessive caloric intake the body stores excess energy in the form of lipids. Over an extended time period this can lead to obesity, which has been linked with type 2 diabetes. Both types of diabetes are characterized by a metabolic confusion, in which key cells reject glucose, while producing more of the sugar via gluconeogenesis, as well as ketone bodies.

Further reading

'Does Thinking Really Hard Burn More Calories?' **https://www. scientificamerican.com/article/thinking-hard-calories**: exploring why mental exertion is hard work.

Nutrition: **http://www.who.int/nutrition/en**: a multimedia repository of nutritional information from the World Health Organization.

'Archibald Edward Garrod and Alkaptonuria: "Inborn Errors of Metabolism" revisited.' **https://www.nature.com/articles/gim201076**: background on Archibald Garrod.

'Why We Just Can't Stop Eating.' **https://www.cam.ac.uk/ cantstopeating?utm_medium=email&utm_source=EN0818**: recent studies into obesity carried out at Cambridge University.

Discussion questions

5.1 Suggest a number of reasons why some people struggle to eat a balanced diet.

5.2 Explain why it is important for the body to maintain levels of glucose within a narrow range in the blood stream.

5.3 Discuss the importance of characterizing insulin resistance.

6 SOLVING TOMORROW'S PROBLEMS WITH NATURAL PRODUCTS

In this chapter, we will explain what natural products are and will discuss their role as primary and secondary metabolites within cells. We will find out how natural products from plants and microbes have become essential components of modern society and look at the potential of these molecules to address grand societal challenges, such as food security and human disease (see Figure 6.1). We will also discover how scientists have developed technologies to produce large amounts of natural products in the laboratory and to modify them. Finally, we will consider some of the ethical issues associated with these scientific breakthroughs.

Figure 6.1 Crocodiles (a) and fungi (b) may appear to have little in common—but both have been used as potential sources of natural products, ranging from antibiotics to food.

(a)

(b)

Anthony Short.

What are natural products?

Perhaps the most useful definition of a natural product was provided by the outstanding scientist, the biochemist and geneticist, Professor Albrecht Kossel (see Case study 6.1). In December 1910, Kossel's life work on the genetic material in cells was recognized when he was awarded the Nobel Prize for Physiology or Medicine. In his introduction speech, as he arrived to collect his award, he announced that:

> *A natural product is any molecule that is produced by a living organism. The study of the living organism has more and more led to the view that its smallest independent units . . .—the cells—. . . are the real seats of the vital processes.*

> *Albrecht Kossel's Nobel Prize Introduction*
> *Speech, December 10, 1910*

As Kossel had discovered, many molecules are produced naturally within the cells of living organisms, including plants, animals and microbes. Natural products can be relatively simple molecules such as urea (also known as carbamide), which has the chemical formula $CO(NH_2)_2$. We now know that these simple molecules are often relatively straightforward to produce commercially—but 200 years ago it was all very different. In 1828, Friedrich Wohler's discovery that he could synthesize urea from ammonia and carbon dioxide in the laboratory was revolutionary. Until that moment vitalism ruled: people believed that life could not be explained by chemical and physical means alone—that there was something special that separated life from inanimate objects. Wohler's synthesis of urea was the first time that a substance previously known as a bio-product of life was created *in vitro*, and it changed the way scientists lookedat the world for good. These days, all the urea used in industry is synthesized artificially, and more than 90 per cent of the urea produced is used as a nitrogen-release fertilizer to improve crop yield in response to global demand for increased food production.

Natural products also include more complex molecules, some of which have been covered in this book, such as the nucleotides required in many areas of metabolism, including for synthesis of DNA and RNA, which you discovered in Chapter 3.

Many natural products have such complex chemical structures that they are extremely difficult, if not impossible, to synthesize in the laboratory, at least at the present time. As far back as 1962, scientists were screening natural products in the hope that they could find a molecule that could be used to fight cancer. They were in luck: a sample originally taken from a scraping of Pacific Yew bark showed great promise (Figure 6.2). However, it took almost another decade to discover, purify, and characterize the active biochemical compound that had these highly sought-after anticancer properties. Paclitaxel (trade name Taxol), as this new pharmaceutical product was called, is a complex natural product (Figure 6.2 (a)). At that time, the extraction and purification methods

Figure 6.2 A potent cancer drug, Taxol (a), was first identified from the bark of a Pacific Yew tree (b).

(a)   (b)

(a): iteranttrader/Wikimedia Commons; (b): inga spence/Alamy Stock Photo.

used by scientists yielded disappointingly small quantities, but by the mid-1980s enough compound was available to allow clinical trials to be well under way. By the mid-1990s, the results of these clinical trials were promising: paclitaxel inhibited the growth of several tumours found in ovarian and breast tissues and Taxol was approved as a chemotherapy drug to treat ovarian and breast cancer.

As was the case with Taxol, one of the problems with many natural products is that those with the most interesting or useful biological activities are only produced in small quantities by cells. This means that it can be difficult to purify enough of these molecules to meet the demands of modern societies. The potential complexity of natural products also presents significant challenges for scientists, who are hoping to synthesize these molecules in the laboratory using traditional organic chemistry methods.

Case study 6.1
Albrecht Kossel

Ludwig Karl Martin Leonhard Albrecht Kossel (Figure A) was born in Rostock, Germany, in 1853, and left home at eighteen years old to study medicine in the University of Strasbourg (which was in Germany at that time). He passed his German Medical License exams in 1877. Kossel was attracted to laboratory research and worked as a research assistant under the supervision of Felix Hoppe-Seyler, the Head of the Biochemistry Department at the University of Strasbourg. It was while working as a research assistant that Kossel discovered nuclein, a biochemical substance that had a protein and a non-protein component. Although little was known about this non-protein component at the time, it was nucleic acid, which contains the genetic information found in all living cells (see Chapter 3).

In fact, it was Kossel who first characterized the four **nucleobases** that form DNA: adenine, cytosine, guanine, and thymine. Guanine had previously been named for where it had first been discovered—in the excrement of seabirds known as guano. Kossel named adenine after the pancreas gland ('adenas' in Greek), which is where it was originally discovered, thymine because it was isolated from DNA purified from the thymus gland of a calf, and cytosine because it was discovered from cells found in calf thymus tissue (and the Greek prefix for cell is 'cyto'). The last nucleobase, uracil, which is used as an alternative to thymine in RNA, was discovered in 1901 by one of Kossel's students, Alberto Ascoli.

He went on to discover a number of amino acids, and a drug to help people affected by asthma and COPD, his son became a well-known physicist, and many of his students also went on to become well known in their fields.

Figure A Albrecht Kossel, who first characterized and named the four nucleobases that form DNA

Archive Pics/Alamy Stock Photo.

Bigger picture 6.1
Biopiracy

During their search for anticancer medicines in the 1960s, the National Cancer Institute screened more than 35,000 plants for any natural products that showed potential to treat cancer. When Arthur Barclay, a botanist working for the US Department of Agriculture, sent a sample of bark that he had chipped off an old Pacific Yew tree to the US National Cancer Institute he had little idea that his plant sample contained molecules that would become one of the most successful drugs of all time; paclitaxel or Taxol is still one of the most highly prescribed drugs for treating cancer. However, the success of Taxol as an anticancer drug was not without its problems. The Pacific Yew tree is a slow-growing tree, only found in old-growth forests located in the northwest Pacific region of America. As excitement about this new drug started to grow, these old yew trees were felled by logging so that their bark could be stripped and enough of the chemically active natural product, paclitaxel (Taxol), could be purified for analysis in the laboratory and for testing on animal models. It took at least three to six Pacific Yew trees to be harvested to treat one cancer patient because the drug could only be produced in limited quantities and this had a negative impact on the forest environment.

As demand for this exciting new drug grew, scientists recognized that they could not rely on Pacific Yew tree bark to meet this demand. They began to search for alternative ways to produce this natural product, but its biochemical complexity meant that chemists could not synthesize this new drug in the laboratory. Instead, they developed semi-synthetic ways to produced it from precursor chemicals found in the needles of a related but much more common (and easy to grow) tree, *Taxus baccata* or the Common Yew tree. These days, scientists can also produce Taxol renewably, from culturing plant cells in the laboratory, producing up to 50 grams of the drug per kg of plant cells.

Many of the drugs we rely on today have their origin in plants, and any active compounds purified from plants using industrial technological approaches are described as **phytopharmaceuticals.** In fact, the World Health Organization estimates that up to 25 per cent of medicines are plant-derived. Although Taxol was found in the bark of trees from ancient forests in North America, prospectors from industrialized countries are looking further afield for the next big discovery. For example, the Amazon rainforest is thought to contain as many as 80,000 plant species, accounting for more than 20 per cent of the known plant species across the globe. However, less than 2 per cent of these plants have been tested for pharmacological activity.

As our demand for new natural products with the potential to solve grand global challenges (such as food security and antimicrobial resistance) increases, attention has turned to the plant kingdom and its natural products. **Indigenous populations** who live in these remote forests often hold vital information about the traditional way that plants have been used in the past to treat certain diseases, and this has provided information that gives us a much more efficient screening method for finding new treatments than relying on chance alone. This ethnopharmacological approach depends on the traditional knowledge held within ethnic groups and, as you might expect, the potential of the Amazon rainforest has attracted the attention of drug companies.

There are examples of plants that have been collected from the Amazon forest and yielded biologically active chemicals that have been patented in other, mainly western, countries and used commercially across the globe without the knowledge of the indigenous populations. This practice of exploiting naturally occurring biochemical or genetic material for commercial gain without offering fair compensation to the indigenous community from which it originates is known as **biopiracy**. It is not unique to South America; in fact it occurs across the globe. Pharmaceutical and biotechnology industries (usually from developed countries) exploit less developed countries, as the indigenous populations often do not receive any benefit from the profits of the foreign companies.

Governments across the globe have responded to this issue with the introduction of the Nagoya Protocol. This agreement sets out to ensure that there is:

> the fair and equitable sharing of the benefits arising from the utilization of genetic resources, including by appropriate access to genetic resources and by appropriate transfer of relevant technologies, taking into account all rights over those resources and to technologies, and by appropriate funding, thereby contributing to the conservation of biological diversity and the sustainable use of its components.

> The Nagoya Protocol, October 2014

Figure A The noni fruit of Costa Rica is used by local people to 'cure' everything from cancer to the common cold. Now scientists are investigating whether any of the many chemicals it contains are biologically active. Costa Rica has signed the Nagoya protocol—so if the fruit does contain therapeutic chemicals, everyone should benefit.

Anthony Short.

? Pause for thought

More than 100 countries have signed up to the Nagoya Protocol since it was agreed in October 2014, including Costa Rica (Figure A). However, there are still significant concerns about whether this agreement is fit for purpose. What do you think these concerns might be?

Primary and secondary metabolites

'Metabolism' is the term used to describe the coordination of enzymatic reactions that take place in living organisms, as described in Chapters 4 and 5. Natural products are produced during these enzymatic reactions and they can be divided into two major classes: the primary and secondary metabolites. Primary metabolites are found in all cells and they are fundamental to the structure and physiology of living organisms. These primary metabolites are essential for cell survival as, without their regulated production, cells and organisms would die. These products include the carbohydrates, proteins, fatty acids, and nucleotides that are the building blocks of all cells (see Chapters 1–3).

On the other hand, secondary metabolites are not essential for survival but they do increase the competitiveness of the organism within its environment, and as such they have been selected for through natural selection. These secondary natural products have many functions. They include:

- the perfumes and pigments that attract insect pollinators to plants and enhance fertilization;
- toxic chemicals produced by plants that ward off herbivores (including microbes, insects, and mammals), protecting the plants from attack;
- chemicals produced by certain plants to enhance their ability to survive in drought or high salt conditions;
- compounds produced by bacteria that are involved in extracellular signalling to other microbes in their vicinity (quorum sensing);
- chemicals produced by animals, fungi, plants, and bacteria which are toxic to pathogens.

The production of secondary metabolites is controlled at the level of gene expression. Cells have evolved complicated and elegant systems to ensure that the genes that produce the proteins and enzymes that synthesize these natural products are tightly controlled. This ensures that these secondary metabolites are produced as and when they are required and the cell does not invest energy in synthesizing products that offer no benefit to the cell or the organism at a given moment in time.

Of course, this natural larder of secondary metabolites offers incredible possibilities. We humans already use these products as food flavourings, dyes, fragrances, pharmaceuticals (including phytopharmaceuticals), and insecticides (see Table 6.1).

Table 6.1 Examples of secondary metabolites and natural products that have been used by humans

Compound	Use	Source	Chemical structure
Vanillin	Vanilla flavouring in food	*Vanilla planifolia* or the Vanilla Orchid grown in warm temperate climates like Mexico and Madagascar	
Colchicine	Medication most commonly used to treat gout	*Colchicum autumnale* or the Autumn Crocus	
Carmine	Red dye used as a food colouring (Cochineal) and as a natural fabric dye, and in lipsticks	*Dactylopius coccus* or the Cochineal beetle found in South America	
Quinine	Treatment for malaria caused by the *Plasmodium falciparum*	*Cinchona officinalis* tree found in South America	
Menthol	Fragrant essential oils, which is the main aroma of peppermint	*Mentha Spicata* or peppermint	Menthol

Biotechnology

In its broadest sense, biotechnology is considered to be the use of biological processes, organisms, or systems to generate products that are intended to improve the quality of human life. Biotechnology is often associated with scientists (see Case study 6.2), big pharma, and genetic engineering. However, the earliest biotechnologists were people who used yeast to raise their bread dough or ferment their drinks, and farmers who developed and improved species of plants and animals by incremental small changes using cross-fertilization and cross-breeding over thousands of years. Biotechnology has also led to the industrial production of natural products, such as alcohol and acetone from plant-based sugar and starch molecules.

Case study 6.2
Károly Ereky

Károly Ereky was born in Esztergom, Hungary in 1872 and he is regarded by many as the father of biotechnology. He graduated from the University of Budapest in 1900 with a degree in technical engineering, and by 1919 he had become the Hungarian minister for nutrition. He was a prolific writer and published over 100 papers, but perhaps his most influential work was the book he published in Berlin in 1919, called *Biotechnologie der Fleisch-, Fett- und Milcherzeugung im landwirtschaftlichen Grossbetriebe* or *The Biotechnology of Meat, Fat and Milk Production in an Agricultural Large-Scale Farm*. In this book, he described how technology could be used to convert raw materials into more useful products. Ereky foresaw how this concept could provide solutions to many 'grand challenges' that still cause societal problems today such as food and energy shortages. For Ereky, '**biotechnologie** (or **biotechnology**)' is the process of taking raw materials and using biological processes to upgrade them into socially useful products. In the early twentieth century, his influential ideas had extended across the globe, and governments in Germany, The Netherlands, Britain, Australia, and Canada could see the value of this approach.

Sadly for Ereky, his influential ideas were also one of the causes for his own demise. After the end of the Second World War, and shortly after the Nazi Germans left Hungary, thousands of civilians fell victim to the Soviet regime that occupied Hungary and needed people who could be used as forced labourers. In addition, there was a drive to find war criminals who may not even have committed a crime during the war but were still considered a threat to rebuilding the country as a modern democracy. In 1946, Ereky was arrested and sentenced to twelve years hard labour by the People's Tribunal, after being found guilty of 'his counter-revolutionary role' in public life. He died in captivity at the age of seventy-four in 1952 and was buried in an unmarked grave. But, thanks to his vision, and almost a century after his seminal book was first published, biotechnology has led to a new industrial revolution.

The manipulation of bacterial biochemical processes can also be exploited to extract precious or expensive metals from low-grade ores. This process, called bioleaching, uses bacterial enzymes to break down the ores, allowing the precious metals to be released and extracted. Bacterial biochemistry can also be used to clean up sites that have been contaminated from industrial or manmade waste in a process known as bioremediation. Although this contamination is harmful to humans and other animals, certain bacteria can use the contamination as a source of energy or as building blocks for cell growth and development.

Biotechnology relies on either the exploitation of biochemical processes, as seen in bioremediation, or the synthesis of a useful end product, such as ethanol. In nature, the biochemistry that underpins both the process and the creation of natural products depends on key protein molecules known as enzymes (see Chapter 2). The production, regulation, and activity of enzymes are dependent on the genetic material that encodes for these vital proteins and this provides significant possibilities for scientists. Hundreds of thousands of different enzymes are involved in secondary metabolism in plants. In fact, there are instances where the synthesis of multiple products can be catalysed by a single enzyme, either from different substrates or from an identical substrate.

For thousands of years, people have had to rely on painstaking and time-consuming cross-breeding to select for the genes that code for the proteins that form the enzymes that deliver the positive traits found in livestock and crops. To a significant extent, cross-breeding relies on chance to generate a desired genetic outcome. However, this process cannot be tightly controlled or regulated, and, as genes from both parents are shuffled during fertilization, the desired genetic outcome can never be guaranteed. Even if a desired trait is successfully selected, there is always the possibility that another unidentified but potentially harmful trait may also have been acquired as part of the breeding process. Nevertheless, cross-fertilization, breeding, and selection have served us well for thousands of years (see Figure 6.3).

As of 2018, scientists have sequenced more than 35,000 genomes from bacteria, eukaryotes, and viruses. If we focus on plant genomes alone, analysis of the genome sequences that we have at the current time suggests that they contain 20,000–60,000 different genes, and perhaps 15–25 per cent of these genes encode enzymes used as part of secondary metabolism. It goes without saying the potential benefits that natural products from these systems may offer society is tantalizing. The discovery of novel natural products with the potential to treat disease or enhance the nutritional value of our food is a distinct possibility, and it is easy to see why pharmaceutical companies, the biotechnology sector, and the food industry have tried to patent the genomes of plants and microbes because of the potential profits available. Before 2013, 4300 human genes alone were patented, but the Supreme Court in the US determined that DNA is a product of nature and so genes cannot be patented. However, genetically modified organisms and sequences used in manipulation **can** be patented and this has been done thousands of times for organisms including maize, rice, and soybeans.

As our scientific knowledge and understanding has advanced, and technology has improved dramatically, we have gained access to a wide range

Figure 6.3 Boran cattle (a), Watusi cattle (b), Devon reds (c), and Holstein/Friesian cows (d) all look different as a result of selective breeding over many years around the world

(a)   (b)

(c)   (d)

(a): isabel hutchison/Alamy Stock Photo; (b): Eric Nathan/Alamy Stock Photo; (c): Dick Kenny/Shutterstock.com; (d): Henk Bentlage/Shutterstock.com

of procedures that can be used to modify or engineer living organisms for human benefits. Scientists are able to use molecular biology techniques and gene cloning to alter genomes by adding or removing pieces of DNA that code for genes that can change the phenotype of the organism that has been manipulated. For example, the combination of restriction enzymes and DNA ligase has allowed gene sequences to be moved between DNA molecules in genetic modification. The expression of the genes has been changed in novel ways, or even in different organisms. This process of genetic engineering is much faster than traditional plant breeding. It has another advantage because the laboratory techniques that scientists now have at their disposal are so advanced that they are able to use exquisite precision to make sure only the gene(s) of interest are moved from one genome to another, preventing the likelihood of other undesirable genes being moved into the new organism. They can also specify the exact site of gene insertion within the receptor genome, reducing the risk of detrimental phenotypic traits, using new technologies such as CRISPR-Cas9 (see Scientific approach 6.1).

One of the first examples of a genetically engineered product involved a human gene that was added into a bacterium, *Escherichia coli* (*E.coli*), in 1978 by a biotechnology company called Genentech. The human gene encodes for a natural product; the hormone insulin (Figure 6.4), a peptide hormone produced by beta cells of the pancreatic islets. As you saw in Chapter 5, lack of insulin causes the life-threatening condition, diabetes (see also the Oxford Biology Primer *Hormones* by Hinson and Raven). Before insulin was produced by genetically engineered bacteria, it had to be extracted from the pancreas of animals from the abattoir. The advantage of genetically engineered insulin is that large amounts of it can be produced cheaply under controlled laboratory conditions, giving a reliable and plentiful supply of pure human hormone.

Once the insulin gene is inserted into *Escherichia coli*, the genetically modified organisms are grown in liquid media in large tanks called fermenters (Figure 6.5). The bacteria replicate quickly, doubling their numbers every thirty minutes or so until a sufficient quantity of bacteria is available in the vat to start to synthesize insulin. Until this point, the bacteria are prevented from making insulin by a repressor protein that sits near the insulin gene. Insulin production is switched on when a chemical, called an inducer, is added to the bacteria. The insulin gene is exposed and transcribed, allowing the bacteria to synthesize insulin. After a few hours, the bacteria are harvested using centrifugation to spin them into a pellet at the bottom of a vessel. The liquid media is removed and the bacterial cell membranes are broken down, allowing

Figure 6.4 The chemical structure of insulin, showing six protein molecules coming together to make the active form in humans

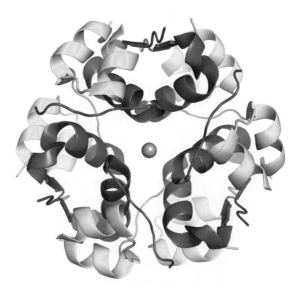

petarg/Shutterstock.com.

Figure 6.5 Biotechnology has had a major impact on insulin production

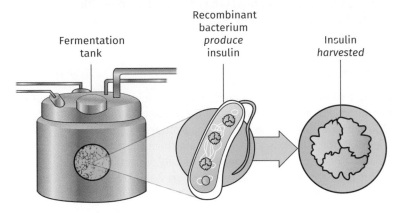

Courtesy U.S. National Library of Medicine.

Scientific approach 6.1
Biofortification: plant breeding versus genetic engineering

It goes without saying that plants are essential components of our diet. Plants provide carbohydrates, fibre, lipids, and proteins, essential natural products that include vitamins, and many essential minerals. But plants also provide **phytonutrients**, compounds that prevent disease. Phytonutrients include polyphenols, sterols, carotenoids, and polyunsaturated fatty acids (see Chapter 2). Looking back at human history, our genome and cellular metabolism evolved when humans were hunter-gatherers and had diets that were rich in fruit, vegetables, and protein but had relatively little fat and carbohydrates. Over the past 10,000 years humans began to cultivate and breed cereal crops that were a rich source of starch and sugars and this has culminated in a highly processed western diet that contains lots of fat and carbohydrate but relatively little fruit or vegetables.

Although crop production has improved dramatically, an unforeseen consequence has been that staple crops with high calorific yields are often nutritionally poor, especially in developing countries. This has led to diseases and mortality resulting from a lack of vitamins and minerals in a nutritionally poor, plant-based diet. In western and developed countries, there is a different problem. Despite public knowledge about the benefit of plant-based food (e.g. five portions a day, as stated in the UK), there is still a reluctance to adopt a plant-rich diet. This has also led to a nutritionally poor diet. Attention has focused on taking staple crops, or plant-based foods that are commonly consumed, and improving their nutritional value so that they produce more vitamins and minerals in a process called **biofortification**. Plant biotechnology has played and can play an important role in engineering natural products like vitamins or other natural products with proven health benefits to biofortify crops so that they supply adequate essential nutrients to prevent the occurrence of deficiency diseases. A number of processes have been developed. These include:

- **Gene enhancement**

Anthocyanins are water-soluble natural products normally found in the leaves, flowers, and fruit (including berries) of plants. These compounds have a red, purple, or blue colour. In flowers, the anthocyanins attract animal pollinators and in fruit they attract animals to come and eat them, aiding seed dispersal. There is also evidence that these molecules can protect plants against environmental stressors, such as cold temperatures and UV radiation. It has been shown that when these compounds are included in the diet of cancer-prone mice they can extend lifespan by 30 per cent, and there are other suggestions that they support good cardiovascular health. This suggests that anthocyanins are important health-promoting natural products. Scientists led by Professor Cathie Martin at the John Innes Centre (Figure A) have used genetic engineering to move genetic material from snapdragons into tomato plants, and this facilitates the production of a purple pigment pathway in the tomato plants to encourage them to produce an increased level of anthocyanin. The result is intense purple tomatoes (see Figure B),

Figure A Cathie Martin, a British plant geneticist, who led much of the work on gene enhancements and anthocyanins in crops

Andrew Davis, JIC photography.

which are incredibly rich in these healthy natural products. The aim of targeting tomatoes for biofortification is that they are a staple product, even in highly processed western-based diets—think pizza and ketchup! Juice from these purple tomatoes is currently being tested to examine whether the anthocyanin has health benefits in humans.

- **Genetic engineering**

Another approach to biofortifying staple crops is to enhance the concentration of provitamin compounds that they contain. Vitamin A or retinol cannot be synthesized by the human body and we rely on our diet to provide this vital natural product. We need vitamin A to maintain healthy eyes and good vision, and to help fight infections. Foods that contain vitamin A include liver, milk, eggs, and fish-liver oils. Another substance called **beta-carotene** (provitamin A), which is found in green leafy and orange/yellow vegetables and fruits, can also be converted to vitamin A. In parts of the world where societies are surviving on a subsistence diet, lack of vitamin A and the blindness that results is a significant issue. Scientists have used biofortification to improve the level of provitamin A by increasing the concentration of beta-carotene in staple crops such as rice and sweet potatoes. Scientists have genetically engineered golden rice (see Figure B) by taking the genes that code for enzymes involved in the synthesis of beta-carotene from the bacterium *Erwinia uredovora* and maize and inserting them into the genome of rice, which did not have these enzymes, to produce a crop that can contain up to 50 per cent of the recommended daily allowance for provitamin A in the diet. This is potentially life-changing and even life-saving in many countries—the World Health Organization estimates that 250,000–500,000 children become blind as a result of vitamin A deficiency every year—and about half of those children then die within a year of becoming blind.

- **Crop breeding**

Glucoraphanin is a natural product found in broccoli. When we eat broccoli, we convert the glucoraphanin into sulforaphane, which is undertaken by the beneficial bacteria that live in our digestive systems. A few hours after eating broccoli, sulforaphane is found in our blood stream before it enters the liver and other tissues. Scientific studies suggest that this compound can reduce the risk of developing chronic heart disease and cancer. Scientists have used a programme of cross-breeding to develop a super broccoli, Beneforté, which is rich in glucoraphanin. Scientists have used field trials in Europe and North America to test Beneforté broccoli to make sure that it consistently yields high levels of this important natural product. In addition, this broccoli has been bred to be much more effective at absorbing sulfur from the soil. This is important because sulfur is an essential component of glucoraphanin. The production of Beneforté broccoli was the result of crossing a high-glucoraphanin broccoli relative from Sicily with cultivated broccoli in the 1980s and, after several years of plant breeding, field trials, and studies into the health benefits, Beneforté broccoli can now be found in supermarkets (see Figure B).

Figure B Biofortified foods such as those shown here have the potential to impact on health and well-being in both developing and developed countries

(i)   (ii)   (iii)

Purple tomatoes   Golden rice   Beneforté broccoli

(i): Andrew Davis, JIC photography; (ii): JIANG HONGYAN/Shutterstock.com; (iii): Seminis/Beneforté Broccoli.

the insulin to be released. The insulin is purified from the bacterial proteins by separation chromatography before being packaged for human consumption.

The biotechnology of insulin production has continued to advance. In 1987, a genetically modified yeast *Saccharomyces cerevisiae* was used to produce human insulin. The advantage of using yeast is that these organisms export insulin directly into the broth, which simplifies the downstream purification process. Recombinant human insulin has also been successfully expressed and produced in the oil seeds of the model plant *Arabidopsis thaliana*. The oil seeds can be easily harvested and the insulin separated using fewer chromatography steps.

The demand for insulin is likely to grow; the World Health Organization estimated that insulin sales will grow from $12 billion to $54 billion globally over the next few years, as diet and lifestyle choices are causing a dramatic increase in the incidence of type II diabetes. Looking forward, there is a real concern that current biotechnological manufacturing technologies will not be able to meet this growing demand due to limitation in production capacity.

Quirky biochemistry

You would be forgiven for thinking that the basic building blocks of life are only sugars, lipids, nucleotides, and amino acids. Using the instruction manual found in the chromosomes of each living cell, these simple building blocks come together in almost endless varieties of sequences and three-dimensional structures to create the biochemical structures and reactions required to sustain all life forms, including microbial cells. There is more about these essential building blocks of cells in Chapters 1, 2, and 3. A central dogma has been established that it takes a triplet of nucleotide bases to code for a single amino acid, and amino acids build the large peptides that we know as proteins and enzymes. The instructions that dictate the order of amino acids that make each individual protein is provided by our genes; our genes code for proteins.

As marvellous as these molecules are, there are natural products made by microbes that are even more interesting. Antibiotics are natural products that have changed modern medicine. Since Alexander Fleming discovered penicillin in 1928, we have relied on these drugs to combat infectious diseases (that can be life-threatening) caused by bacterial pathogens. The vast majority of antibiotics are natural products made by microbes, including soil bacteria. Antibiotics are quirky molecules that have a passing resemblance to the conventional building blocks of life. Just like most other biochemical structures and reactions, antibiotics rely on a suite of genes tucked into the genomes of the microbial cells that produce them. These genes code for the production line of enzymes that work together to synthesize each individual antibiotic.

Many antibiotics are natural products and they come in a wide variety of different shapes, sizes, and structures. These natural products are produced by soil bacteria, including *Streptomyces*. Although the syntheses of some of these natural products have a passing resemblance to conventional protein biochemistry, there are significant and surprising differences. One of these classes of antibiotics is the thiazolylpeptides (the first one, micrococcin, was discovered more than fifty years ago). Although they are not clinically useful (because they are not very soluble), they are effective antibiotics that are able to knock out specific types of bacterial pathogens that depend on a thin cell wall, including the well-known pathogen methicillin-resistant *Staphylococcus aureus* (MRSA), which is responsible for many life-threatening, hospital-acquired infections. They are encoded by short sequences of DNA that are translated into short chains of amino acids called peptides that are forced to undergo a series of serious chemical modifications. A significant section of the front end of the peptide is removed and the smaller peptide molecule left behind, and the set of chemical modifications change it from a floppy linear peptide into a robust, rigid, antibiotic molecule.

Perhaps the most well-known antibiotic is penicillin, first discovered by Alexander Fleming in 1928, as a soluble compound produced by the fungus *Penicillium notatum*. This antibiotic is clinically relevant, because it destroys human pathogens such as *Staphylococcus*. These antibiotics work by interfering with the rigid cell wall that maintains the integrity of the bacterial cell. Without a rigid cell wall, bacterial cells suffer osmotic stress; the cells burst open, causing the bacterium to die. Like the thiazolylpeptides, penicillin starts life as a short sequence of amino acids, but unlike thiazolylpeptides, penicillin is much, much smaller. It begins its synthesis as three single building blocks, two of which are conventional amino acids used in protein molecules. The third, however, is an unusual molecule. It is an unfinished, half-formed amino acid. These three amino acids are stitched together, tweaked, and chemically modified to create the penicillin molecule that has changed modern medicine. Once again, the enzymes involved in the synthesis of this quirky, life-saving molecule are encoded by a suite of genes found side by side in the fungal genome. Although antibiotics have dramatically improved healthcare outcomes against infectious diseases, problems are increasing due to the emergence of bacteria that have resistance to current antibiotics (see Bigger picture 6.2). You can find out much more about infectious diseases and how we treat and prevent them in Chapter 5 of the Oxford Biology Primer *Human Infectious Diseases and Public Health* by Dr William Fullick.

Bigger picture 6.2
Antibiotic resistance

Not only has the discovery of antibiotics such as penicillin and the thiazolyl-peptides changed modern medicine, they have also changed our understanding of the intricate biochemistry that is part of microbial life. Penicillin soon became the go-to miracle drug for treating infections during the Second World War. However, it did not take long before microbial resistance to this drug started to develop. This was the golden era of antibiotic discovery and scientists began to search the planet for other microbes with the ability to produce new and exciting antibiotics. They were in luck. A whole raft of new antibiotics were discovered and pharmaceutical companies invested in the drug discovery and production pipeline.

Although antibiotic resistance continued to emerge to each new antibiotic that was discovered, there was an optimism that new antibiotics would continue to be found. This did not last long and the antibiotic pipeline began to dry up. The same antibiotics were discovered, and rediscovered, again and again. Our reliance on antibiotics as miracle drugs means that demand for them continues to grow and it could be argued that we did not treat these drugs with the respect that they deserved. At the same time, levels of antibiotic resistance have continued to increase, and pharmaceutical companies have stopped investing in the antibiotic drug discovery pipeline. It has been projected that if nothing changes, if no new antibiotics become available for clinical use and our reliance on antibiotics stays the same, then by 2050 antibiotic resistance will be responsible for more deaths than either heart disease or cancer. Society is facing an era that it thought it had left behind, when bacterial infections again become the leading cause of death and amputations across the globe.

 Pause for thought

Clearly one of the most effective ways to address this looming global crisis of antibiotic resistance is to invest in drug discovery and the antibiotic pipeline. Suggest reasons why new antibiotics are not appearing as rapidly as we need them.

Chapter summary

- Natural products are molecules produced by enzymatic reactions that take place in living organisms. They can be divided into two major classes: the primary and secondary metabolites.

- As scientists' understanding of biochemistry develops and technologies advance, we are starting to exploit natural biological processes, organisms, or systems to generate products that are intended to improve the quality of human life by improving health and food security.

- The potential complexity of natural products presents significant challenges for scientists who are hoping to synthesize these molecules in the laboratory using traditional organic chemistry methods, and they also raise ethical issues when we try to purify them from the natural environment.

Further reading

hthttps://www.who.int./news-room/fact-sheets/detail/natural-toxins-in-food: useful online resource about natural products that may be toxic to humans.

https://www.cbd.int/abs/becoming-party/default.shtml: useful online resource about ratification of the Nagoya Protocol.

https://quadram.ac.uk/superbroccoli: online resource describing the development of Beneforté broccoli.

Eaton, S. B. and Konner, M. (1985) 'Paleolithic nutrition. A consideration of its nature and current implications', *New England Journal of Medicine*, 312: 283–9.

Martin, C. and Li, J. (2017) 'Medicine is not health care, food is health care: plant metabolic engineering, diet and human health', *New Phytol*, 216, 699–719 (doi:10.1111/nph.14730).

Ramadan, H. A. I. and Redwan, E. M. (2014) 'Cell factories for insulin production', *Microbial Cell Factories*, 13: 141.

http://doi.org/10.1186/s12934-014-0141-0

Discussion questions

6.1 Industrial manufacturers are much more inclined to describe their products as 'natural' and 'organic' as opposed to 'chemical', 'biochemical', or 'semisynthetic'. Why do you think that might be?

6.2 What do you think that society can do to combat the growing problem of antibiotic resistance?

7 SOLVING TOMORROW'S PROBLEMS: BIOENERGY AND THE ENVIRONMENT

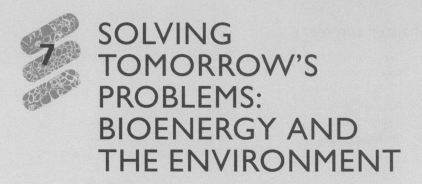

By harnessing our human understanding of biochemistry we have the potential to tackle the environmental problems we face. Climate scientists now agree that human activities are causing the Earth's environment to heat up. Activities—from burning fossil fuels to feeding an ever larger global population—are releasing carbon dioxide and other greenhouse gases, trapping infrared radiation inside the Earth's atmosphere and heating it up.

Biotechnology could help us to reduce our crippling dependency on fossil fuels. Controversy has raged over the science of climate change. Even if climate scientists are wrong in their predictions (and the overwhelming consensus is that they are right!), no one can disagree that alternatives to fossil fuels will be necessary when they eventually run out. Biotechnology, or biotech, which simply means technology derived from living systems, has already been used to pioneer several different forms of renewable energy. Algae, a diverse group of plant-like organisms, have sparked great interest with their potential to produce carbon-neutral biofuels. Also researchers have been developing artificial photosynthesis, which could also be used to produce alternatives to fossil fuels.

Agriculture could also reduce its carbon footprint with help from biotechnology. This field has the potential to produce artificial meat and dairy products, which would reduce our dependence on animals, along with greenhouse gas emissions. Successful prototypes have already been produced and companies are gearing up for their commercial release. Meanwhile, plants have been genetically modified for greater efficiency, to grow in harsh climates and generally to alleviate the scarcity of food that characterizes a snowballing global population (Figure 7.1).

Figure 7.1 The cultivation of oil seed rape has become very common in China as a source of cooking oil and also biodiesel fuel

Cultura Creative (RF)/Alamy Stock Photo.

Biotechnology and the generation of novel forms of energy

Researchers are currently investigating a number of ways in which bio-technology could solve the world's energy crisis. As humans, we depend on plants to convert energy from the sun into the glucose we need to survive. Moreover, photosynthesis created the precursors to the fossil fuels which have sustained our mushrooming population and technological progress since the industrial revolution. Could such biochemical phenomena be har-nessed to produce the fuels we need directly?

Bioengineering life forms to produce fuels would have twofold benefits. First of all, it would provide a sustainable means to meet our fuel demand, once fossil fuels have run out. Second, it would remove carbon dioxide from the air, so that any combustion of fuels would not increase the level of atmospheric carbon dioxide. Such carbon-neutral solutions could slow down the global warming threatening civilization.

Cultivating microalgae to produce biofuels

Algal biofuels are an excellent example of a promising biotech lead that is unlikely to develop into commercial success.

Algae were a very promising research field for many years. Millions of pounds have been invested in the hopes of yielding a profitable alternative

Figure 7.2 The seaweed you see at the seaside is a form of multicellular macroalgae

divedog/Shutterstock.com.

to fossil fuels. As hopes have been dashed, many companies have either gone bust or focused on other ways to make money from algae.

Algae form a diverse group of photosynthesizing organisms. Some are large enough to see, such as seaweed (Figure 7.2). They are known as macroalgae. Many are single-celled organisms, defined as microalgae, which are the variety that have been investigated for the production of biofuels.

Microalgae produce fatty acids which could be used as biofuels. All algae photosynthesize, meaning they are able to transform carbon dioxide and water into glucose. Most glucose is used either to fuel cellular processes or combined with nitrogen to create proteins. When nitrogen is scarce, some of the sugar is transformed into fatty acids, including omega-3 (see Chapter 2). Laboratory researchers have shown how these fatty acids can be harvested to act as biofuels.

Unfortunately, no one has been able to scale an algal biofuel operation up to the commercial level. In fact, Kevin Flynn, a professor at Swansea University, suggests that it would require ponds three times the area of Belgium to meet the targets originally designated for algal biofuels! Moreover, it would require half of all the fertilizer currently used to farm the whole of Europe. Part of the problem is that the algae all compete for light. As more of them grow, they block the light from each other, limiting the rate of photosynthesis and hence growth. Addressing these problems with genetic modification could be very risky (see Bigger picture 7.1).

Sadly, it seems algal biofuels will not be the boon visionaries once imagined. Luckily, there are other potential sources of renewable energy.

Artificial photosynthesis

Artificial photosynthesis is another option for renewable energy. Such technologies harness the power of sunlight to produce fuels from carbon dioxide and water. While many artificial photosynthesis systems are completely inorganic, one of the most promising research directions involves biotech.

In 2016, a researcher at Harvard University reported a great leap forward for artificial photosynthesis using bacteria. Daniel Nocera, a professor of energy science at the Ivy League university, and his colleague Pamela Silver used a biotech system to produce liquid fuel with an efficiency of 10 per cent. That means that one-tenth of all the light energy they captured was transformed into chemical energy. If that does not sound like much, keep in mind that plants only achieve a 1 per cent efficiency rate. Moreover, 10 per cent was much higher than any other artificial photosynthesis system known at that time.

Nocera and Silver's team achieved their coup with a combination of inorganic and organic methods. First of all, an inorganic cobalt–phosphorus catalyst was used to split water via electrolysis into hydrogen and oxygen. Next, *Ralstonia eutropha* bacteria were recruited to oxidize the hydrogen (Figure 7.3). This microorganism is unusual in its ability to metabolize carbon dioxide and hydrogen to produce poly[(R)-3-hydroxybutyrate], a polymerized methyl ketone (see Figure 7.4). Moreover, teams including Nocera's have genetically modified the microorganism to increase its

Figure 7.3 *Ralstonia eutropha* bacteria have been used to boost inorganic methods of artificial photosynthesis to an efficiency rating of 10 per cent

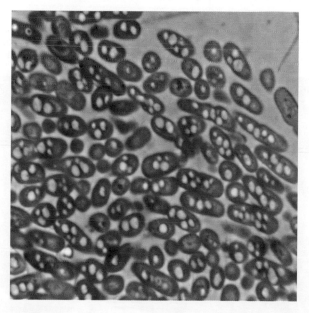

Image courtesy of Christopher Brigham/MIT. http://creativecommons.org/licenses/by-nc-nd/3.0.

Figure 7.4 Structure of poly[(R)-3-hydroxybutyrate], a polymerized methyl ketone produced by *Ralstonia eutropha* bacteria from carbon dioxide and hydrogen

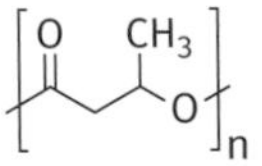

output of the combustible fuel. Artificial photosynthesis could be a boon for renewable energy but it remains to be seen how important the role of biotech will be in this emerging technology.

Faecal fuel

By 2015 in the UK, Brian Harper had had enough of seeing plastic bags of dog poo hanging from trees. He responded by designing Britain's first dog poo-powered street lamp, which now sits proudly in the Malvern Hills area of outstanding natural beauty in South West England (see Figure 7.5). Harper follows in the footsteps of nearby city Exeter, where a sewage sludge digester powered local street lamps as early as 1890. In fact, the generation of power from faeces has an even longer tradition. Since the Neolithic period humans have tapped excrement for energy. To this day some people in India still dry patties of cow dung in the sun for fuelling fires (see Figure 7.6). And in many societies animal and even human faeces are used in small-scale digesters to produce methane for cooking and electricity generation. Manure is a rich source of energy and bacteria help to tease it out.

Harper's 'dog poo lamp' has a very simple design. Dog walkers deposit dog faeces into a hatch. Inside a bioreactor employs microorganisms to break down the organic matter and release methane gas. The process is anaerobic, meaning that oxygen should be absent from the bioreactor. Finally the methane gas lights the lamp.

Harper's design is standard in the biogas world. The system is very straightforward: first gather sewage, manure, or other suitable organic waste. (Many biogas reactors harvest energy from landfill sites.) Next, channel the waste into an oxygen-free bioreactor with suitable methanogenic bacteria. Finally, collect and distribute the resulting methane gas. Pressure

Figure 7.5 The dog faeces-powered lamp developed by Brian Harper, which can be found in the Malvern Hills area of outstanding natural beauty

Nic Fleming.

Figure 7.6 A woman prepares cakes of cow dung to use as fuel

Salvacampillo/Shutterstock.com.

must be maintained within a narrow range of levels because combustible methane gas can explode if the pressure gets too high or too low.

Diverse strains of microorganism are suitable for biogas reactors. Many strains, such as *Methanobacterium bryantii*, are named as bacteria but are actually species of archaea, an ancient and biologically distinct domain of microorganisms.

Some countries make more use of biogas than others. Such findings show that the UK could make a significant impact on its energy demands by making more use of biogas technology.

Biogas boosts sanitation as well as providing energy. In the developing world latrines are not always available, and people often defecate anywhere outdoors. Even when human waste is buried in the ground, it causes infectious diseases such as cholera, while encouraging flies and other insects, which carry pathogens to other unsuspecting hosts. Wells can also become contaminated. As sewage needs to be treated anyway, it makes sense to employ bioreactors which not only reduce the health risks posed by excrement, but also provide energy.

As the energy crisis continues to grip humanity, this affordable biotechnology should increase in popularity.

Fibre as fuel

It's a common fact of life that when you eat sweetcorn, you can see the kernels afterwards in your faeces. The body cannot digest these kernels because they are made primarily of cellulose. As we saw in Chapter 1, pandas also

Figure 7.7 Crops of sugar cane have been used to produce bioethanol from the sugar. New biotechnologies could produce ethanol from the fibrous lignocellulose plant material that is normally thrown away. This would alleviate the problem of sacrificing agricultural land to produce biofuels.

Isarapic/Shutterstock.com.

lack the enzymes to break down cellulose. Even cows do not have the necessary enzymes, but they can digest cellulose thanks to gut bacteria and an arrangement of different stomachs between which they transfer cuds of semi-digested grass. Bacteria such as these are the focus of efforts to extract biofuels from inedible plant matter.

A major drawback of biofuels has been their impact on food supplies. Growing sugar cane to produce ethanol deprives people of food that could have been grown in the same place. But imagine if biotechnology was able to ferment ethanol from the parts of the sugar cane that we cannot digest (see Figure 7.7). This is one of the plans of biotechnologists. Cane stalks contain a lot of fibrous matter, rich in cellulose and another biopolymer called lignin, which is normally thrown away. But if researchers can harness the digestive powers of the bacteria like those found in the stomachs of cows, they may be able to transform this plant waste into bioethanol. Now sugar cane crops would not only provide bioethanol but also sugar, which could be sold in the usual way.

Plants can be grown with softer cell walls to make the process even more efficient. Lignocellulose is a mixture of cellulose and lignin, both of which are very resistant to being broken down, even by bacteria. Two strategies have been employed to reduce the cost of biofuel extraction from this inedible plant matter. The first is to breed plants to make them produce less lignin. The second is to genetically modify plants with cell wall-degrading enzymes from bacteria. In this way the plant attacks its own cell wall with imported enzymes. Research has found that this method can be employed with no reduction in the crop yield. Biofuels can be extracted more profitably from plants with softer cell walls.

Hard-to-reach fossil fuels

Biotechnology could also be invaluable for maximizing returns on fossil fuel drilling operations. Currently, oil companies are accustomed to leaving up to 80 per cent of crude oil in the ground, because some of the deposits are too difficult—and therefore unprofitable—to drain completely (Figure 7.8).

Biotechnology could free up these hard-to-reach deposits by transforming them into methane gas. Microorganisms of the bacterial genus *Syntrophus* and archaeal genera including *Methanosaeta*, *Methanospirillum*, and *Methanoculleus* are able to convert organic compounds such as alcohols and alkanes into methane gas. In fact, this technique could even be combined with carbon capture and storage-style technologies. One potential solution to the crisis of accumulating atmospheric carbon dioxide is to bury carbon dioxide underground. Such deposits could be inoculated with microbial cocktails to transform the carbon dioxide and stubborn fossil fuel deposits into methane gas, which would float up to ground level for collection. Such techniques would reduce the level of atmospheric carbon dioxide, making the process more carbon neutral than simply drilling for fossil fuels.

Figure 7.8 Microorganisms could harvest methane from crude oil, huge proportions of which have to be left underground because they are not profitable to drill

TebNad/Shutterstock.com.

Biotech could also treat fuels to reduce pollution. Microbes can be used to remove impurities such as nitrogen, metals, ash, and sulfur, the combustion of which in coal leads to the formation of acid rain. Microorganisms can also reduce the viscosity of crude oil, lowering the energy demands of refining it, along with the volume of carbon dioxide emissions. However, these initiatives could yet prove problematic at the commercial scale and prove unviable, like the algal biofuels.

Using biotechnology to reduce human impact on the environment

Biotechnology has the potential to reduce our carbon footprint in ways other than by producing novel fuels. Another major source of carbon is agriculture, and particularly cattle farming.

Providing food for an ever more gargantuan global populace produces carbon in a number of ways. Farming vehicles and machinery such as tractors and combine harvesters require energy, the majority of which is still provided by carbon dioxide-releasing fossil fuels.

Meat production, especially of beef, is another problem. Cows are consumers, meaning they have to eat plants. Cattle are often fed soya beans to make them grow faster but growing the soya beans already comes with its own carbon footprint, especially when they are imported from the developing world, where the necessary land is often produced by deforestation. The situation with dairy cows is particularly serious. Cows are large animals which need a lot of food, but the actual proportion of energy required to produce milk is a tiny fraction of their overall energy consumption. Innovators have started to wonder if the dairy milk can be produced without the cow.

Artificial milk

Biotechnology has a proven track record of downsizing the production of valuable substances. Insulin, the drug prescribed to help diabetics regulate their blood sugar levels, used to be extracted from pigs. Nowadays it is produced by genetically modified bacteria, which are substantially cheaper to feed than pigs and produce a reliable supply of pure human insulin. Pioneering firms are now trying to use the same principle to produce dairy milk proteins from genetically modified yeast cells.

Casein and whey, the main proteins found in dairy milk, provide both nutrition and flavour. When we eat proteins our bodies break them down into amino acids, which we use to make our own proteins. This is why our brains reward us when our tongues detect proteins. Our tongues are adapted to detect at least five tastes: salty, sour, sweet, bitter, and savoury. Essential nutrients such as sugar and salt have a pleasant taste so that our ancestors knew the right things to eat. Meanwhile, bitter and sour tastes warn us off potentially harmful foods. Proteins and the amino acids they are composed of activate our savoury taste receptors, which is why replicating the proteins found in dairy milk is such an important part of reproducing the flavour in artificial milk.

Figure 7.9 Biotech pioneers are trying to produce artificial milk by genetically modifying yeast to produce the casein and whey proteins that largely characterize the flavour of dairy milk. This could lead to a whole range of dairy products that taste like the real thing but do not originate from cows.

Anthony Short.

Suppose we wish to produce casein from a genetically modified organism (see Figure 7.9). The first step is to identify the gene, the collection of nucleobases that dictate the sequence in which amino acids will be connected in the protein. Casein is well understood and its gene sequence can easily be looked up in research literature. Step two is to produce a large number of copies of the casein gene, a process we will consider in more detail in 'DNA printing' below. Next, these genes are spliced into the chromosome or plasmid of a suitable single-celled organism, such as yeast cells (Figure 7.10). The modified cells are then placed in bioreactors and provided with suitable nutrients, especially sugars, so that they reproduce and express the casein. Finally, the molecules of casein are released from the cells and collected.

DNA printing

Let's look in more detail at step two. The process of producing copies of a gene is known as DNA printing. First of all, the gene sequence is inputted into a computer. Next, the computer operates a reactor module which automatically connects nucleobases in the desired sequence to simultaneously produce large numbers of copies of the gene.

In the reactor copies of the gene are grown by the sequential addition of nucleobases. At the heart of the process is a special surface, engineered so that nucleobases can form chemical bonds to it. One by one nucleobases are added to the reactor, so that gene copies are grown on the surface and then harvested, not unlike the way wool grows out of a sheep's skin and is then shaved off for use in clothing.

Figure 7.10 Yeast cells can be genetically modified to produce the casein and whey proteins that characterize the flavour of dairy milk

Anthony Short.

Figure 7.11 The nucleoside guanosine has a variety of nucleophilic regions which could bond in the wrong position with neighbouring nucleosides

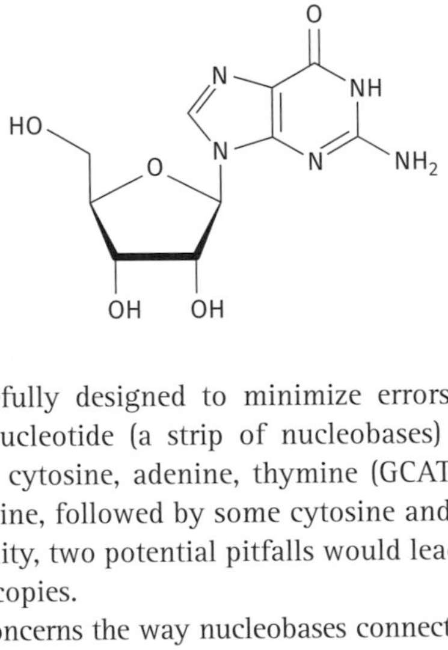

The system is carefully designed to minimize errors. Suppose we wish to make an oligonucleotide (a strip of nucleobases) with the following sequence: guanine, cytosine, adenine, thymine (GCAT). In theory we can just add some guanine, followed by some cytosine and then the remaining nucleobases. In reality, two potential pitfalls would lead to a hotchpotch of mismatching gene copies.

The first pitfall concerns the way nucleobases connect. Look at the guanosine nucleotide shown in Figure 7.11. Each nucleotide should connect via the 3′ and 5′ position hydroxyl groups on the ribose group. However, the 2′ position hydroxyl group and the amine group on the nucleobase also have lone pairs of electrons which could launch nucleophilic attacks on neighbouring molecules. Left unchecked nucleotides could join in the wrong orientation to each other, which would render the genes unable to express the desired proteins.

The other pitfall is that the growing gene copies might skip a nucleobase. Suppose a million guanosine molecules have successfully docked to the reaction surface. Next, the surface is sluiced with cytosine molecules but some of them fail to form bonds to the guanosine molecules. Now a proportion of the growing chains feature lone guanosine molecules, while others feature the desired guanosine and cytosine sequence. If we now sluice the surface with molecules of the next nucleotide, adenosine, some of them will bond directly to the lone guanosine molecules, which now have the potential to grow into mutant chains. Systems are required to remove such truncated genes from the field of play.

The first pitfall is addressed by adapting nucleotides to disable some of their functional groups. Phosphoramidites are special versions of nucleotides that have been modified with protecting groups. Consider again the nucleotide guanosine. A protecting group could be added to its amine group, which would then prevent the amine from reacting with other molecules. Phosphoramidites usually have four protecting groups. Incoming phosphoramidites are only able to form bonds via the 3′ position ribose hydroxyl group. Next, another chemical is added which removes one of the protecting groups from the 5′-position ribose hydroxyl, priming the nucleotide to bond in the correct orientation to the next incoming nucleotide. Once all of the nucleotides have been added, additional chemicals are added to remove the remaining protective groups.

The second pitfall is also addressed with a protective group. In the earlier scenario a million guanosine molecules had moored to the reactor surface, but not all of them had formed bonds to the cytosine molecules added next. At this stage a reagent is added which selectively binds a protective group to the guanosine molecules, but not the cytosine molecules. In this way the truncated genes are removed from play, as the next round of nucleotides— the adenosine molecules—are only able to form bonds to the cytosine nucleotides. In this way malformed copies of the gene are pruned from the process, minimizing waste and ensuring that all of the copies produced are faithful to the requested sequence.

DNA printing is a staple in the world of biotechnology, enabling the transformation of written data into physical gene copies at the touch of a button.

The finished product

The final stage of artificial milk production is to add the remaining ingredients. Having modified microorganisms to produce proteins, the other ingredients have to be supplied from elsewhere. Milk contains a lot of fat, for which plant-based oils can be substituted. Sugar and salt will also need to be added and some other ingredients too. A San Francisco-based start-up called Perfect Day is currently developing artificial dairy milk for commercial release. At the time of going to print they had already produced a successful prototype by genetically modifying yeast cells to produce dairy milk proteins. Next, they will scale up the production process to meet the needs of the mass market.

One of the great advantages of artificial milk is its suitability for people with dietary problems. Many people suffer stomach upsets or skin problems when they consume lactose, the sugar found in dairy milk. As you saw in Chapter 1, the answer concerns epigenetics, the field of genetics which dictates which genes get expressed and when. Humans are generally born with the gene for lactase, the enzyme that metabolizes lactose which is found in breast milk. But in many people this gene is silenced once they stop feeding on breast milk as infants. What this means is that other genes modify their DNA so as to stop the lactase gene from being expressed. Such people will then struggle to digest the sugar and are likely to have problems when they consume dairy products, like milk, butter, and cheese. This gives Perfect Day the edge with their product. There is no reason for them to add lactose when their milk can be sweetened with other sugars. This will open up the delights of cheese pizza, yoghurt, and other delicious milk-based foods to people who suffer with dairy intolerance.

Forging steak

In 2013, a researcher and a food critic sat down to sample a brand-new kind of meal. The two judges delivered a lukewarm verdict on the very first public serving of laboratory grown meat—there was no stampede to buy it!

Artificial meat is another way that biotechnology is poised to shrink our carbon footprint. Many organizations have now produced passable substitutes for meat by culturing stem cells. Early research has tended to focus on dishes made from minced meat, such as hamburgers, since prototypes have been little more than bundles of cells. The great challenge will be to turn two-dimensional layers of cells into three-dimensional slabs of meat.

Stem cells will develop to become more specialized. For example when a baby is first conceived, the embryo is composed of stem cells. During the course of the pregnancy these cells will gradually specialize into specific types, such as neurones in the central nervous system, or muscle cells. In the body such specialized cells cannot turn back into stem cells, but many methods have been found to achieve this in the laboratory. Such cells have great potential to cure debilitating diseases such as Alzheimer's or multiple sclerosis. The technology could also be harnessed to reduce the impact of meat production on the environment.

Early research has focused on a specific kind of stem cell. Specialization is not a binary process, in which a blank canvas becomes a masterpiece in a single step. Throughout their lifespan stem cells will gradually specialize, passing checkpoints of decreasing potential, the term used to describe how free they are to specialize. For example, mesenchymal cells have the potential to become any of the cells involved in skeletal muscle, but could no longer become neurones. The stars of musculature are the striated muscle cells whose individual contractions combine to make the entire muscle contract. Bones and the connective tissues that bind them to muscles are formed of two more important cell types which originate from the partially specialized mesenchymal cells. A final important example, as we shall see below, is adipose cells, which store excess energy in the form of fat deposits.

The plans for this developing technology are to include increasing numbers of the different cell types. Early prototypes were formed solely from striated muscle cells. This name relates to the striped appearance of the cells forming the major muscles, like biceps and quadriceps, as opposed to the smooth muscle cells found in blood vessels and other sites. The first laboratory-grown meat to be introduced to the public, described in the introduction to this section, was produced from striated muscle cells (see Figure 7.12). Following the verdict Dr Mark Post, who led the research, planned to introduce a second type of cell. The flavour of meat is strongly characterized by fat, so the researcher planned to include adipose cells in future laboratory-grown meats.

Diversifying cell types should also lead to improvements in texture. When stem cells are cultured in the laboratory, they are grown in petri dishes along with all the necessary nutrients for the cells to thrive. In these conditions it is very difficult to grow a sheet of tissue with a thickness of more than one cell. The reason is that the outer cells tend to absorb all the nutrients, leaving nothing for those in the middle. In the body this problem is overcome with the vascular network, the blood vessels which deliver nutrients and oxygen to cells throughout the body. In order to go from mincemeat to steak, researchers will have to find a way of replicating the function of blood vessels. The obvious suggestion is to induce the mesenchymal stem cells to form still another variety of cell—the kind that forms blood vessels. In this way, cells could be cultured ever more closely in form to actual musculature, with the potential to transform artificial minced meat into artificial steaks or chicken breasts, and so on.

Figure 7.12 Professor Mark Post presents the world's first ever artificially cultured hamburger, grown from bovine mesenchymal stem cells

Mosa Meat.

The great hope is that biotechnology will allow people to continue enjoying meat without damaging the environment. It should be possible to tune the technology to produce meat using less energy and also releasing a lower mass of the greenhouse gases associated with cattle farming, namely carbon dioxide and methane.

Feeding our food

While algae have been a disappointing source of fuel, the organisms could still address the growing problem of food security. Kevin Flynn, the professor mentioned in the section 'Cultivating microalgae to produce biofuels', argues that algae can be farmed as a food source to cultivate fish and shellfish, as well as to produce essential fatty acids for human consumption, such as omega-3 fatty acids needed for heart health.

Invaluable information can be salvaged from the disappointing research into algal fuel. As referenced above, many of the difficulties of producing algal fuels centred around the difficulty of cultivating algae in sufficient numbers. While it now seems impossible to cultivate algae on a big enough scale to produce biofuels profitably, this expertise can be transferred to its cultivation for agricultural purposes.

We not only have to feed ourselves, but also our food. Earlier in the section we saw how crops of soya beans are grown as feed for cattle. Commercially farmed fish such as salmon also need to eat and they tend to be fed fishmeal, which is becoming an unsustainable practice.

Fishmeal is essentially ground up fish (Figure 7.13). When nets catch smaller fish that people generally do not eat, they can be ground up and fed

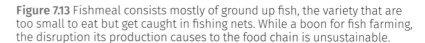

Figure 7.13 Fishmeal consists mostly of ground up fish, the variety that are too small to eat but get caught in fishing nets. While a boon for fish farming, the disruption its production causes to the food chain is unsustainable.

Doidam 10/Shutterstock.com.

to cattle or to commercially farmed stocks of fish. So while farming salmon may appear to be a sustainable way to reduce the pressure on our disastrously over-fished oceans, it's an own goal if they need to be fed fishmeal. Fishmeal also includes portions of ground up bones and other things which are thrown away from fish prepared for a human market—in making fillets of fish for example. Fishmeal is, in one way, a good use of waste, but the use of fishmeal for fish-farming puts an unsustainable pressure on the oceans.

Algae could help to reverse some disturbing oceanic trends. Roughly one-third of the carbon dioxide released since the industrial revolution has dissolved into the oceans. Once dissolved, the gas reacts with water to produce carbonic acid (H_2CO_3), which lowers the pH of the ocean. The increasing acidity of the ocean is harming aquatic life. For example, oysters farmed in the US often die before they grow out of the larval stage. As we have seen, algae are able to photosynthesize, meaning they absorb carbon dioxide. In other words, they could counteract some of the harmful effects of the rising concentration of carbon dioxide in the sea. Meanwhile, where many species are struggling, algae will thrive in the changing conditions of the ocean, including its rising temperature.

Algae can be an excellent source of nutrition. Humans rely strongly on oily fish to provide omega-3 and omega-6 essential fatty acids (Figure 7.14). In fact, these vital nutrients are also essential for the fish. That is, many fish are unable to make their own omega-3 and -6 fatty acids. So while humans obtain their essential fatty acids from fish, the fish ultimately obtain them from algae, at the very bottom of the sea-life food chain. Algae can also provide all of the essential amino acids. As such algae could be used to produce a sustainable alternative to fishmeal.

Figure 7.14 Tuna fish is a rich source of omega fatty acids for humans but tuna themselves obtain the essential fatty acids from algae via smaller fish

(a)   Linoleic acid

(b)   α-linolenic acid

(c)

(c): Shane Gross/Shutterstock.com.

Figure 7.15 Seaweed has formed an important part of the human diet for millennia. Harvesters living on Pearce Island, British Columbia, traditionally lay out squares of the macroalgae while waiting for the community's fishermen to return.

Amy Deveau.

Algae could even become a food source for humans. In fact, many cultures already eat certain species of algae: seaweed (Figure 7.15). But there are many challenges ahead for the incorporation of algae into the human diet. First, mass-producing algae is no easier when producing food as it is when making biofuels. Second, getting the nutrition out of algae and into humans is far from straightforward. Not all algae produce the same nutrients. In fact, even the same species will produce different nutrients depending on where they are, what season it is, and other factors. Third, no one can be sure what will happen if we eat algae. Just because a person eats algal cells, there is no guarantee that the nutritional payload will be transferred into our digestive season. Moreover, as microorganisms they might influence the activities of our gut bacteria, or even the epithelial tissue lining our gut. In a worst-case scenario, such biochemical interactions could contribute to the development of colon cancer. Much research is still necessary to identify which algae to eat, how to cultivate them, and how to extract their nutritional cargo in a healthy way.

Genetic engineering could help to reap the benefits locked up in this ancient life form. Studies have already investigated whether algae can be genetically modified to boost their output of valuable nutrients, such as fatty acids. Algae could also be modified to make more efficient use of light during photosynthesis or to spend more time photosynthesizing. But many people believe this would be a risky course of action (see Bigger picture 7.1).

Bigger picture 7.1
GM or no GM? The importance of informed debate

Echoes of the past chime in a breakaway strand of the genetic modification debate—this time around microalgae.

Around the turn of the century, the public became very engaged with the potential for genetic modification to inflict serious, irreversible damage on the ecosystem. Many different people took part, some better informed about the science than others. Well-known and influential people, as well as a number of celebrities, expressed opinions which were not always backed up by science—but how can the average person in the street know who is expressing evidence-based information and who is expressing an opinion that—however deeply felt—may be ill-informed?

People listen when celebrities condemn new technologies. Many public figures warned that once organisms had been tinkered with, the changes could replicate themselves throughout nature with devastating consequences. Popular culture has a much wider audience than almost any scientist, so this message really caught the public's attention (Figure A).

In spite of this outcry, genetic modification is now common practice. Many supermarket foods are produced from genetically modified crops, especially

Figure A The media storm that developed around the idea of genetically modified foods when they were first introduced has been negatively influencing public thinking, and even Government policy, ever since it began, in spite of all the scientific evidence for the potential global benefits genetic modification of crops can bring

Glenn Copus.

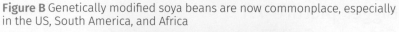

Figure B Genetically modified soya beans are now commonplace, especially in the US, South America, and Africa

Bill Barksdale/Agstockusa/Science Photo Library.

soya beans (Figure B), even if the product itself is not directly genetically modified. Meanwhile, bacteria were genetically modified several decades ago to produce insulin, which has had amazing benefits for people affected by diabetes around the world. And as we have seen, yeast can be modified to express dairy proteins, which can be used to make artificial milk. It is easy to object to genetic modification living in a wealthy, developed country with plenty to eat. The voices of those dying through lack of food, or lack of nutritionally rich food, who would benefit greatly from the benefits GM crops bring, do not grab the headlines in the same way.

In spite of these examples, some people are still strongly opposed to the genetic modification of microalgae. The worry is that genetically modified algae could run rampant, spreading 'alien' genetic material around the globe.

In 2017, the campaign group Friends of the Earth condemned the genetic engineering of microalgae. Speaking in response to a study carried out by Sapphire Energy, in partnership with the University of California at Los Angeles, the pressure group said commercial operations involving genetic modification of algae should not be allowed. The study in question deliberately investigated whether modified algae would spread into the wild. Sapphire Energy had modified the microalgae to produce more fatty acids but the focus of the study was to check whether it would spread. They reported that their modified algae did indeed spread, but did not overtake wild-type strains of the organism. However, Friends of the Earth interpreted the findings to show that GM microalgae could indeed spread into the wild. Their spokesperson Dana Perls likened it to a genie that could never be put back into the bottle.

What triggered the revival of these historical objections to genetic modification? In the simplest terms, any important decision should be based on the context. It depends on the purpose of the modification, which organism will be modified, and how the organism will subsequently be cultivated.

The colossal demand for fuels dictates the likely modifications to microalgae. As mentioned in the section *Cultivating Microalgae to Produce Biofuels*, it would require lakes three times the size of Belgium to make enough biofuel to meet demand. In other words, the algae would have to be cultivated in colossal volumes, which would be impossible to contain from the outside world. Two drawbacks with this are that the algae could be eaten by predators and that they could easily transfer to the environment. One way to reduce the risk of predation would be to modify the algae to contain some kind of poison. Moreover, the area of water required for cultivation could be reduced if the algae were modified to photosynthesize more effectively. Professor Kevin Flynn, the expert in algae from the University of Swansea (Figure D), is a strong opponent of such genetic modification. In this case, the more effectively the algae is modified, perhaps the greater is the risk of problems arising.

If history has taught us anything, it is the danger of introducing species with no natural predator. The highly poisonous cane toad was introduced to Australia as a means to biologically control the cane beetle (*Dermolepida albohirtum*) and French's beetles (*Lepidiota frenchi*), which were ravaging sugar cane crops. Nothing could eat the cane toad in this new environment because no species had evolved to cope with its lethal cocktail of poisons. Consequently the ecosystem was overrun with toads instead of bugs. Arming algae against predators by modifying it to produce toxins would improve its ability to take over ecosystems, encouraging harmful phenomena like algal blooms.

Modifying microalgae to photosynthesize more efficiently presents similar hazards. If algae can produce more fatty acids with less sunlight, it might be possible to meet biofuel targets with lakes twice instead of three times the size of Belgium. But as studies have already shown, these GM breeds could infiltrate wild-type algae, a set of species famously able to spread around

Figure C An early biotech success was the production of medicinal insulin from genetically modified bacteria. This has revolutionized treatment for people affected by diabetes. Previously, insulin was extracted from cows or pigs, which resulted in an irregular supply of the vital drug, allergic reactions for some patients, and a problem for some religious groups.

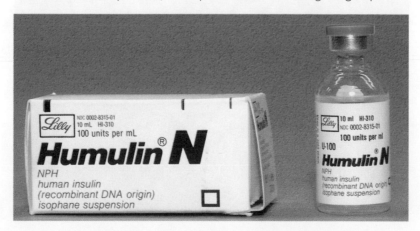

Division of Medicine and Science, National Museum of American History, Smithsonian Institution.

the globe like a pandemic virus. These GM algae could quickly take over algal colonies with their superior photosynthesis, enabling them to chisel a much deeper foothold into global ecosystems, just like the cane toad (Figure E) but on an even bigger scale.

Figure D Professor Kevin Flynn, an expert in algae based at the University of Swansea, has voiced grave concerns about the genetic modification of openly cultivated microalgae, along with doubts about its capacity to meet energy demands

With permission from Professor Kevin Flynn.

Figure E The introduction of cane toads as a means of biological control to Australia was disastrous. With no natural predators, the highly poisonous amphibian took over

Anthony Short.

This logic does not apply to all genetic modification. Modifying yeast and bacteria carries a much lower level of risk. The genetically modified organisms are grown in containers—the huge fermenters do not give them access to the outside world. If such microorganisms find their way into the environment, there is less danger they will spread. For example, the bacteria modified to produce insulin (see Figure C) are unlikely to find an adaptive benefit of producing insulin in the wild. Meanwhile, they will have to find food. Contrast this with genetically modified algae. The very reason that researchers are interested in microalgae is because it can photosynthesize. If these genetically modified strains find their way into the wild, they will not run out of food because they make their own.

Algae are renowned for spreading rapidly. Unlike genetically modified crops, they grow in aquatic media in three dimensions, can move from one place to another, and reproduce at a very high rate. They are spread in many different ways, including in ship's ballast and by birds. Algae can quickly become problematic, for example when run off from fertilizer feeds they rapidly grow in ponds. Soon the pond can be overrun with algae, which blocks out the light for the species beneath. The water plants die, and decomposers use oxygen from the water as they break down the dead plants. Herbivores starve as there are not enough plants to feed on, then carnivores too—and again oxygen is used up as they decompose. Soon all life has gone, apart from the algae at the top (Figure F). Imagine if a genetically engineered form of algae took to the wild, how much more rapidly it might form such algal blooms if it had been modified to photosynthesize more efficiently, while also repelling predators with its human-programmed poisons.

Figure F Algal blooms form when fertilizer is leached into aquatic environments. The algae block out the light, with lethal consequences for the aquatic life below the surface.

Tom Archer/MI Sea Grant.

On the other hand, using genetically modified algae to produce food is a different proposition. Algae are already being cultivated commercially to produce omega-3 supplements. The manufacturers keep the algae in bioreactors. Not only is this a necessary condition of producing a foodstuff, to prevent contamination, but it also prevents the spread of algae to wild communities.

Unfortunately, many people do not stop to differentiate between the potential use of genetically modified algae in the oceans to produce biofuel, with its many possible problems, and the use of genetically modified algae in bioreactors to produce food and dietary supplements.

Biotech still plays a vital role in algal cultivation, even if the organisms are not genetically modified. Biochemical expertise guides the manufacturing process to maximize productivity. For example, if a company wished to cultivate algae to produce fatty acids, it would be vital that they starved the algae of nitrogen for a short period of time directly prior to harvest. Why is this? Algae have to be manipulated to produce fatty acids. Like humans, the microorganisms convert surplus glucose into fatty acids, but for different reasons. We use it to store energy, as we saw in Chapter 4. It is adaptive for algae to transform excess glucose into fatty acids to prevent the sugar from leaching out of their cells, where it might feed bacteria, which could then colonize the surface of the algae. Normally, any glucose left over from respiration would be combined with nitrogen to make proteins. But if nitrogen is withheld, by stopping the flow of fertilizer, the algae will switch from protein to fatty acid synthesis. All of this is invaluable expertise for any company cultivating algae to produce fatty acids.

As in so many scientific debates, the hazards of genetic modification relate to responsible usage. There are certain scenarios where genetic modification appears safe, desirable, and potentially part of the solution to many of the problems we have generated for ourselves—and others where we should proceed with caution. Ideally, international guidelines will be developed and adhered to so that everyone can benefit, and risks are kept to a minimum all over the world. And education—so everyone in the population understands the science behind the debate and is less vulnerable to emotive and ill-informed claims for or against genetically modified organisms—is a key factor going forward.

⬚ Chapter summary

- Biotech could help to meet the energy demands of the modern age but many initiatives are in the embryonic stages and may fail at the commercial scale.
- Doubts have arisen over the capacity of microalgae to meet the colossal demand for biofuels. The microorganism can be cultivated to produce fatty acids, which can be converted into effective fuels, but it seems doubtful they can meet demand for the quantities necessary.
- Biotech could also help to harvest energy from inedible fibrous plant material or convert abandoned crude oil deposits into readily

collectible methane gas. Its use to harvest natural gas from organic waste is expected to grow in popularity.

- One of the most promising uses of biotech is the production of artificial foodstuffs. Artificial dairy milk and meat prototypes have already been produced and will likely go on sale in the next few years.

- Concerns still remain over the use of genetic modification. Although the practice is now commonplace, the issue is context specific. Doubts have been expressed over the safety of modifying microalgae to produce biofuels. Modifying microalgae for more effective cultivation could also equip the species to more effectively devastate wild areas, which it might infiltrate.

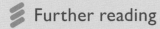 Further reading

'The Seaweed Site: Information on Marine Algae' **http://www.seaweed.ie/algae-l**: a guide to the diverse species of algae.

'China's Appetite Pushes Fisheries to the Brink'

https://www.nytimes.com/2017/04/30/world/asia/chinas-appetite-pushes-fisheries-to-the-brink.html: an exploration of the sustainability challenges surrounding the use of fishmeal.

'Future Applications of Biotechnology to the Energy Industry'

https://www.ncbi.nlm.nih.gov/pmc/articles/PMC4741079: a more detailed investigation of nascent applications of biotechnology in the energy industry.

'Algal Biofuel Production is Neither Environmentally Nor Commercially Sustainable'

https://theconversation.com/algal-biofuel-production-is-neither-environmentally-nor-commercially-sustainable-82095: Kevin Flynn's argument about the challenges facing microalgae as a source of biofuel.

Discussion questions

7.1 To what extent can each of the biotech energy initiatives be considered carbon neutral? Should we continue research into the cultivation of biofuels or focus on technologies that eliminate the release of carbon dioxide altogether, such as solar?

7.2 How could scientific information be used to allay concerns consumers might have about eating artificially cultured foodstuffs, like lab-grown meat and artificial milk?

7.3 How does context influence the debate over whether or not to adopt genetic modification of different organisms?

GLOSSARY

α-helix Conformation of a polypeptide backbone consisting of a rigid, rod-shaped helix formed when the N–H group of one amino acid forms a hydrogen bond with an oxygen of a COOH group of an amino acid that is four residues away. The side chains of the amino acids point outwards from the helix, meaning small, uncharged amino acids are likely to be involved in this type of structure.

acid Molecule (or ion) capable of donating a proton (hydrogen ion, H^+) as part of a chemical reaction.

activated carrier A biomolecule that can transfer a group of atoms or electrons to another molecule in an energetically favourable process.

adenine Purine base found in all types of nucleotides and nucleic acids.

adenosine-5′-triphosphate (ATP) Nucleotide that acts as a universal currency of energy in biological systems, in addition to being a precursor for RNA and DNA.

aerobic Biochemical processes or reactions that require oxygen or occur in its presence.

alkaptonuria Metabolic disorder caused by problems with enzyme homogentisate-1,2 dioxygenase which turns urine black and damages joints and heart valves.

amino acid One of 20 different, naturally occurring molecules, each with a unique shape and biochemical properties. Amino acids have the same overall structure, with a central carbon atom attached to: a basic amino group, which can accept a proton to become positively charged (NH_3^+); an acidic carboxyl group, which can lose a proton to become negatively charged (COO^-); a hydrogen atom; one of 20 different chemical groups, usually referred to as the 'R' group or side-chain.

amphipathic Molecules that contain hydrophobic (water-hating) and hydrophilic (water-loving) regions.

anabolic reactions Biochemical pathways and reactions that synthesize larger biomolecules from smaller compounds; generally these require energy.

anaerobic Biochemical processes or reactions that occur in the absence of oxygen.

anoxygenic Biochemical processes by which certain types of bacteria use light to consume carbon dioxide without release of oxygen.

antibiotics Antibacterial agents suitable for use to treat infection, having good selective toxicity. Historically, the term was reserved for naturally produced agents, but it is now commonly used also to include synthetic and semi-synthetic molecules.

anthocyanins Water-soluble, natural products normally found in the leaves, flowers, and fruit of plants.

archaea Single-celled organisms that are fundamentally different to bacteria. These were among the first free-living cells on Earth.

atomic number Number of protons in the nucleus of an atom.

ATP synthase Protein complex that synthesizes ATP from ADP and inorganic phosphate.

β-pleated sheet Conformation of a polypeptide when two (or more) sections of the chain line up side by side and hydrogen bonds form between the polypeptide backbone and the oxygen of COOH groups of adjacent chains. They can be parallel, where the polypeptide chains run in the same direction, or antiparallel, where the polypeptide chains run in opposite directions.

base Molecule (or ion) capable of receiving a proton (hydrogen ion, H^+) as part of a chemical reaction. In aqueous solutions they often release hydroxide ions (OH^-). **In relation to nucleic acids:** the term base (or nucleobase) refers to the nitrogen-containing compounds that are an important component of nucleotides.

beta-carotene A precursor of vitamin A, found in green leafy vegetables and in orange/yellow vegetables and fruits.

bioenergetics Studies of energy flow through living systems.

biofortification Altering plants to produce more vitamins and minerals in order that they have improved nutritional value.

bioinformatics Interdisciplinary science that combines biology, computer science, information engineering, and statistics to analyse and interpret biological data.

bioleaching Process using bacterial enzymes to break down metal ores.

biopiracy Practice of exploiting naturally occurring material for commercial gain without offering fair compensation to indigenous communities.

bioremediation Processes that use microorganisms to break down pollutants to clean polluted materials or sites.

biotechnology Use of molecular biology and genetic engineering methods to generate biological molecules that can be applied to solve specific challenges or problems to society.

carbohydrates (also called saccharides) Molecules that are made from just three elements—carbon, hydrogen, and oxygen—with the general molecular formula $(CH_2O)_n$.

catabolic reactions Biochemical pathways and reactions that break down larger biomolecules into smaller compounds; generally these make energy available for cells.

centromere Part of a chromosome that links a pair of sister chromatids.

chiral molecules (also see stereoisomers) Molecules that are linked to different chemical groups, leading to asymmetry. Chiral molecules are not identical to molecules that are their mirror image.

chromatid One half of two identical copies of a newly formed chromosome. The two chromatids are joined by a single centromere to make a chromosome.

chromatography Laboratory methods that separate different components of a mixture. Different biophysical properties may be used, such as size, charge, or shape.

chromosome DNA-protein complex containing part (or all) of the genetic material of an organism. In these complexes, the DNA molecule takes up less three-dimensional space than when it is unbound to proteins.

circadian rhythm Change in the level of a biomolecule (or the effect it produces) that is dependent on the time of day over an approximately twenty-four hour cycle.

cloning The production of genetically identical individuals or molecular biology techniques that alter genomes by adding or removing pieces of DNA that code for genes.

coding DNA DNA bases present within a genome that provide the instructions to synthesize specific protein sequences.

condensation reaction Reaction forming a covalent bond between two molecules, leading to loss of a molecule of water from the starting molecules.

conjugate acid The chemical group formed when a base receives a proton during a reaction. (The conjugate acid always has one more H atom than the base.)

conjugate base The chemical group formed after acid donates a proton to a reaction. (The conjugate base always has one less H atom than the acid.)

covalent bond (also called a molecular bond) Chemical bond that involves the sharing of electron pairs between two distinct atoms.

crassulacean acid metabolism (CAM) Plants with this metabolism keep the stomata in the leaves shut during the day to reduce evaporation, but open at night to collect CO_2.

cytosine Pyrimidine base found in all types of nucleotides and nucleic acids.

deoxyribonucleic acid (DNA) Nucleic acid providing hereditary material for most cells. The sugar for each nucleotide is 2′-deoxyribose and the sequence of bases provide genetic instructions for all biochemical processes occurring in the cell. It usually occurs as a two-stranded helix.

dimer Molecule consisting of two similar types of monomer.

DNA ligase Enzyme that joins specific ends of DNA strands together by forming a phosphodiester bond. An essential, final reaction during many processes involving DNA, including replication.

duplex Double-stranded.

endosymbiosis Process by which a host cell—an original prokaryote—absorbs another prokaryotic cell to the mutual benefit of both.

energy transduction Biochemical descriptions of the way energy is transferred between different molecules.

enthalpy Energy transfer that occurs during a chemical reaction, usually defined by H.

entropy Extent of randomness or disorder, usually defined by S.

enzyme Biomolecule that acts as a catalyst for a chemical reaction, enabling it to happen with enough precision and speed to maintain life.

essential biomolecules Biomolecules that cannot be synthesized by the host organism, so they must be taken in via the diet.

fatty acid Long chains of hydrocarbons, typically varying in length between twelve and twenty carbon atoms. Each fatty acid contains a methyl group (–CH$_3$) at one end and a carboxyl group (–COOH) at the other end. If the carbon chains contain only C–C single bonds they are described as saturated fatty acids, those with a single C=C bond are described as monounsaturated, and those with two or more C=C bonds are referred to as polyunsaturated.

fermenters Large vessels filled with liquid media used to grow microorganisms on an industrial scale.

free energy Amount of energy available during a reaction (or process) that is available to do work.

gene Section of a DNA that encodes a specific biological function, usually through synthesis of a specific protein.

genetic engineering Molecular biology techniques that alter genomes by adding or removing pieces of DNA that code for genes.

genome Full complement of genetic information present in each cell.

glycogen phosphorylase One of the enzymes that breaks down glycogen during periods of high glucose demand, e.g. exercise.

glycogen synthase The enzyme that adds molecules of UDP-glucose to the growing chain of glycogen during glycogen synthesis.

glycogen synthase kinase The enzyme that deactivates glycogen synthase during periods of high glucose demand, e.g. exercise.

guanine Purine base found in all types of nucleotides and nucleic acids.

homeostasis The maintenance of a dynamic equilibrium within a narrow range in the body, e.g. body temperature, blood glucose levels.

hydrolysis reaction Reaction that breaks a covalent bond, leading to the addition of a molecule of water to the products.

indigenous community Groups of people that have historical claims to be the original settlers of a given geographical region.

ionic bond Occurs when one atom completely gives up one (or more) electron(s) to another atom.

isomer Two (or more) molecules with the same chemical formula but a different arrangement of atoms in the molecule. Often isomers will have different chemical properties.

isotopes Atoms with the same atomic number but different numbers of neutrons.

ketone bodies An alternative fuel source comprising acetoacetate, d-3-hydroxybutyrate, and acetone, which the body produces when there isn't enough glucose available.

kinetic energy Energy of an object due to its movement.

lipid A diverse group of molecules that perform different biochemical roles in living organisms. They are often relatively insoluble in water, but they are all able to dissolve in a non-polar solvent.

l-selenomethionine A dietary supplement in which sulfur is substituted for the trace element selenium in the amino acid methionine.

mass number Number of protons plus the number of neutrons in the nucleus of an atom.

metabolism (metabolic pathways) Sum of all biochemical processes that take place within a cell (or organism).

micro RNAs (miRNAs) Small RNAs, typically about twenty-two bases long, that regulate expression of specific genes.

molecular biology Science studying the roles of individual molecules in biological processes. Often used to refer to the range of laboratory methods that allow cloning and manipulation of DNA sequences.

monomer Small molecule that is the building block of a biological polymer. In biology, these are usually amino acids, nucleotides, and monosaccharides.

natural product Any molecule that is produced by a living organism.

nanotechnology Technological solutions to problems, usually involving manipulation of specific molecules.

non-coding DNA DNA bases present within a genome that do not provide direct instructions for protein sequences. Often these sequences are important for regulation of gene expression.

non-coding RNA RNA molecule that is not translated into a protein. Different types of non-coding RNAs occur, including transfer RNAs, ribosomal RNAs, and a range of small RNAs and long, non-coding RNAs.

non-essential biomolecules (e.g. fatty acids) Biomolecules that can be synthesized by the host organism.

non-polar Chemical groups or bonds where electrons are shared equally and there are no dipoles.

nucleic acids Biological polymer made from chains of nucleotides. See also deoxyribonucleic acid (DNA) and ribonucleic acie (RNA).

nucleobases The bases found in DNA and RNA.

nucleoside Similar structures to nucleotides, but containing only a base bound to a sugar.

nucleotide Chemicals found in all cells, with the same three general components: a nitrogenous-containing base, a sugar, and one (or more) phosphate groups. When present in RNA and DNA, the sugars are ribose or 2′-deoxyribose, respectively.

oligomer Molecule consisting of a small number (typically between three and ten) of similar types of monomer.

oxygenic Biochemical processes (in plants and certain types of bacteria) that use light to consume carbon dioxide, with release of oxygen.

papillae Bunch of sensory nerve cells that are on the surface of an organ, e.g. taste buds on the tongue.

periodic table Grouping of chemical elements by their atomic number. The elements are divided into different rows (periods) and columns (groups).

phenotype Set of observed characteristics of an organism resulting from the interaction of its genotype with the environment.

phenylketonuria A medical condition characterized by the failure of the body to metabolize the amino acid phenylalanine.

phosphodiesterases Enzymes that break a phosphodiester bond in molecules such as cAMP, cGMP, or stretches of DNA or RNA.

photosynthesis Biochemical process that exploits the energy from sunlight to synthesize organic molecules.

phytonutrient Natural products present in plants that are required for their growth.

phytopharmaceutical Medically active compound purified from plants using industrial technological approaches.

polar Chemical groups containing atoms with different potential to attract electrons, leading to an electric dipole moment, with negatively and positively charged ends. Such groups are able to form hydrogen bonds and easily interact with water.

polymer Large macromolecule made of lots of identical (or similar) monomers. In biology, these can be proteins, nucleic acids, or polysaccharides.

polymerase chain reaction (PCR) A cyclical process that makes large amounts of specific DNA sequences using a DNA polymerase that is able to withstand high temperatures. It is widely used in forensic sciences and gene cloning technologies.

polysaccharides Biological polymer made from chains of saccharides.

potential energy Energy of an object due to its position relative to other objects.

primary metabolite Natural products that are fundamental to the structure and physiology of each living organism.

primary structure The sequence of monomers present in a polymer. Commonly used for proteins to refer to the sequence of amino acids dictated by the genetic information encoded by the gene.

proline One of the naturally occurring amino acids, with a closed ring structure between the side chain and amino group.

proton-motive force Energy generated by the transfer of protons across a membrane which is used for chemical, osmotic, or mechanical work in cells.

protein Biological polymer made from chains of amino acids.

purine One of two types of base found in nucleic acids. Each purine contains two carbon–nitrogen rings, with the most common types being adenine and guanine.

pyrimidine One of two types of base found in nucleic acids. Each pyrimidine contains a single carbon–nitrogen ring, with the most common types being cytosine, thymine, and uracil.

quaternary structure Three-dimensional structural arrangement of all chemical groups in proteins composed of more than one poly-peptide chain, or additional ligands that are not part of the primary sequence. The multiple chains, or ligands, are held together by hydro-gen bonds, hydrophobic interactions, or covalent cross-linking.

receptor Mechanism used by cells to moni-tor and respond to changes in the environment. Most receptors are proteins.

recombinant DNA Novel DNA molecule prepared using gene cloning, containing DNA sequences that are not present in nature.

replication Biochemical process that synthe-sizes a new DNA molecule, copying the sequence of bases from both strands of the original DNA molecule.

repressor Molecule (usually a protein) that inhibits the expression of one (or more) genes by binding to DNA or RNA associated with the gene.

respiration Process that makes biochemical energy available due to the oxidation of organic compounds, typically with the intake of oxygen and the release of carbon dioxide.

restriction enzymes Enzymes that cut DNA molecules at (or near) a specific sequence of bases. They have been exploited in genetic engineering.

ribonucleic acid (RNA) Various forms of this nucleic acid are involved in different aspects of protein synthesis, allowing a cell's genetic information to be converted into specific cel-lular structures or activities. The sugar for each nucleotide is ribose.

saccharides See carbohydrates.

secondary metabolite Natural products that are not essential for survival of an organism, but they increase its competitiveness within its environment.

secondary structure Three-dimensional struc-tural arrangement of chemical groups in part of a polymer. Most commonly used in proteins to refer to specific conformations of the poly-peptide backbone, such as the α-helix and the β-pleated sheet.

somatic cell Any body cell in an organism which is not a reproductive cell.

standard atomic weight (also called the relative atomic mass) Average mass of the full range of isotopic mixtures of an element found on Earth.

stereoisomers Molecules containing the same atoms, but where their three-dimensional spatial arrangement means they are mirror images that cannot be superimposed over each other.

synthetic biology Science that uses prin-ciples from engineering to take advantage of and improve biological systems, usually with appli-cations to tackle specific problems.

telomere Region of repetitive bases at the end of a linear chromosome. These protect the chromosome end from shortening and deteri-oration and from joining with neighbouring chromosomes.

tertiary structure Three-dimensional con-formation of atoms found in a specific polymer. Commonly used in proteins to refer to the inter-actions between all amino acid side chains.

thermodynamics Studies of biochemical pro-cesses that describe the relationship between all forms of energy in a system.

thermodynamic parameters A system and its surroundings.

thymine Pyrimidine base. Mostly found in deoxyribonucleotides and deoxyribonucleic acids.

total energy The sum of the potential and kinetic energies of a system.

translation Biochemical processes converting gene sequence information into proteins.

transcription Process in which RNA polymerase uses the DNA sequence of a gene to make an RNA copy, specifically referred to as messenger RNA (mRNA).

triglycerides (also referred to as triacylglycerols) Molecule with a fatty acid attached via an ester bond to each of the three carbons in glycerol. Most molecules contain a mixture of different fatty acids, which can have different lengths, varying levels of saturation, and different physicochemical properties. Triglyceride mixtures are referred to as fats or oils: fats are solid at room temperature and oils are liquid.

uracil Pyrimidine base. Mostly found in ribonucleotides and ribonucleic acids.

urea (also known as carbamide) Main nitrogenous breakdown product of protein metabolism in mammals, eventually excreted in urine. Has the chemical formula $CO(NH_2)_2$.

INDEX